Improving Transparency, Integrity, and Accountability in Water Supply and Sanitation

Improving Transparency, Integrity, and Accountability in Water Supply and Sanitation

Action, Learning, Experiences

María González de Asís
Donal O'Leary
Per Ljung
John Butterworth

The World Bank Institute
and
Transparency International

THE WORLD BANK
Washington, DC

ISBN: 978-0-8213-7892-2
eISBN: 978-0-8213-7867-0
DOI: 10.1596/978-0-8213-7892-2

Improving transparency, integrity, and accountability in water supply and sanitation / María González de Asís . . . [et al.].
    p. cm.
  Includes bibliographical references and index.
  ISBN 978-0-8213-7892-2 — ISBN 978-0-8213-7867-0 (electronic)
1. Municipal water supply—Government policy. 2. Municipal water supply—Corrupt practices. 3. Sanitation—Government policy. I. Asís, María González de, 1969–
  TD345.I475 2009
  363.6'1—dc22

                                                                    2009000085

Cover photograph: Young woman fetching water from a well in Taroudant Province, Morocco; Julio Etchart, World Bank, 2002

Cover design: Critical Stages

Contents

Boxes

Figures

Tables

Foreword

This manual on *Improving Transparency, Integrity, and Accountability in Water Supply and Sanitation* is the result of a partnership between the World Bank Institute (WBI) and Transparency International (TI). It was developed under the Open and Participatory Government Program at the Municipal Level (known by its Spanish acronym as the GAP Municipal Program). The GAP Municipal Program, managed by WBI since 2000, supports institutional change in local government by helping to design tools to combat corruption. It provides a platform for disseminating knowledge on anticorruption strategies that can be adapted and used by national agencies and municipalities worldwide. Over the years, GAP has supported many training initiatives in Latin America and the Caribbean and in Anglophone and Francophone Africa.

Nowhere do citizens, particularly the poor, feel the effects of corruption more directly than at the municipal level. Corruption calls into question the social contract between citizens and public officials whose duty is to provide vital services. TI's Global Corruption Barometer 2006, a survey of the general public conducted in 62 countries, found that bribery in poor and transitional countries is still a major impediment to development. In Africa, for example, corruption in public services, including utilities, affected more than a third of respondents.

Tackling corruption in municipal water supply and sanitation services requires a holistic approach, focusing on governance reform and particularly on developing and implementing anticorruption strategies at the sectoral and institutional levels. This requires an adequate sector organization that distinguishes clearly between the roles of policy formulation and sector planning, delivery of services and sectoral regulation, access and service quality, and operating efficiency and tariffs and financial performance.

Research on governance has shown that political will—open and unequivocal support at the highest levels—is a prerequisite for anticorruption reform, which can also be strengthened by forming broad-based multistakeholder groups to monitor progress and provide strategic direction. These principles, which underpin the GAP platform, are illustrated in the manual's five modules:

- Module 1 lays out a conceptual framework for understanding the nature of corruption and analyzing the effects of different types of corruption on customers, institutions, and society at large.
- Module 2 discusses how to use internally and externally focused tools to investigate the extent of corruption and the preparedness of service providers and other organizations to prevent it.
- Module 3 presents a suite of tools to address corruption in water supply and sanitation. Many of these tools call for increased participation by civil society organizations in identifying sectoral budget priorities and in monitoring sectoral performance. Improving access to information is a key ingredient in many of these tools.
- Module 4 includes a number of case studies that demonstrate how the use of the tools discussed in module 3 and other anticorruption tools have led to outstanding sectoral and institutional performance in countries as diverse as Cambodia, Panama, and Singapore.
- Module 5 describes how to create and implement action plans to address corruption and improve transparency, accountability, and access to information in the water and sanitation sector.

Although the main focus is on Honduras and Nicaragua, case studies from many regions make the manual universally relevant. Applying the insights and tools described here can help raise the coverage and quality of municipal water supply and sanitation services, thereby contributing to the economic well-being of all citizens and improving civic engagement.

Specifically, this manual can help:

- **Increase the involvement of civil society** by engaging all stakeholders in setting water supply and sanitation priorities and monitoring performance, including reducing opportunities for corruption.
- **Increase the contributions of water supply and sanitation services to poverty reduction** by increasing the quality and coverage of service to poorer communities on an equitable basis.
- **Promote the financial sustainability** of water and sanitation service delivery organizations, thereby increasing the confidence of consumers, civil society organizations, and other stakeholders in those institutions' ability to expand and improve service.
- **Raise ethical standards** among all stakeholders, especially service delivery organizations, thereby instilling a sense of public service throughout these organizations.

Improving Transparency, Integrity, and Accountability in Water Supply and Sanitation is an important contribution to meeting the objectives of the GAP Municipal Program and the larger goal of reducing poverty by improving governance worldwide.

We believe the manual will be of interest to regulators, managers and staff of utilities, consumers (large and small), and contractors as well as civil society organizations.

Sanjay Pradhan   Cobus de Swardt
Vice President   Managing Director
World Bank Institute   **Transparency International**

Preface

The World Bank Institute is one of the World Bank Group's main instruments for developing individual, organizational, and institutional capacity in the Bank's client countries. WBI designs and delivers learning programs that create opportunities for development stakeholders to acquire, share, and apply global and local knowledge and experiences. Through its courses and seminars, professional networks, manuals such as this one, and communities of practice, WBI facilitates South-South and North-South knowledge exchange and innovation.

WBI has identified better governance and the reduction of corruption at the sectoral level as high priorities for improving development effectiveness. Research has shown that poor governance, especially corruption, has proportionally greater adverse effects on those who can least afford it—the world's poor. Governance in the water and sanitation sector is particularly important, not only because access to basic services is an essential element in people's ability to rise out of poverty but also because neglect in service provision can result in devastating declines in public health. High prevalence of diarrhea in children, for example, stems from the failure to treat water properly.

More than 1 billion people live without access to safe, potable water, in part because of poor governance and corruption. Illegal connections and substantial losses caused by deferred maintenance have eroded the revenues of water utilities, leading to a downward spiral in performance. Embezzlement of funds, bribes for access to illegal water connections, manipulation of meter counters, and collusion in public contracts add to the litany of corrupt practices.

We believe this manual is a useful tool for those who wish to diagnose, analyze, and take action against systemic corruption in the water and sanitation sector. Although the manual's central focus is on two countries, many of the principles, case examples, and tools are more broadly applicable. We offer this manual as a practical guide for governments, civil society organizations, and citizens themselves in their quest for a model of service provision that responds to the pressing needs of people in the developing world.

Roumeen Islam, Manager
Poverty Reduction and Economic Management
World Bank Institute

Acknowledgments

The program to improve transparency, integrity, and information in the water and sanitation sector and the resulting manual would not have been possible without the work and efforts of Rafael Vera (WSP-ETWAN), Nelson Antonio Medina (WSP-ETWAN), Gustavo Saltiel (LCSUW), Carolina Vaira (WBIPR), Diana Páez Guajardo (WBIPR), Chris Hale (WBIPR), Isabel Benjumea (WBIPR), Manuel Schiffler (LCSUW), Meike Van Ginneken (ETWWA), David Michaud and Nicole Maywah (LCSUW), Jonathan Halpern (ETWWA), Ana Bellver (LCSPS), Jorge Irias Mena (LCSPS), Alberto Leyton (LCSPS), Maria Emilia Freire (FEU), Charles Kenny (FEU), Shantilall Aatmanand Ramsundersingh (WBISD), Karen Sirker (WBISD), Alejandra Hernández (ETWAN), Mina Lacayo (ETWAN), Ricardo Mairena (ETWAN), and Marie-Luise Ahlendorf (TI).

The valuable support of WBI Vice President Sanjay Pradhan, Acting Vice President Rakesh Nangia, and sector managers Roumeen Islam (WBIPR), Adrian Fozzard (LCCHN), Joseph Owen (LCCNI), François Brikke (ETWAN), Carlos Vélez (LCSUW), and Nicholas Manning (LCSPS) was fundamental for the realization of this program in both Honduras and Nicaragua.

Our local partners in Honduras—the RAS-HON, SANAA, and CONASA (Jorge Méndez, Ligia Miranda, Tupac Mejía)—and in Nicaragua—the CONAPAS (Luis Angel Montenegro), INAA (Carlos Schutze and Benita Ramírez), ENACAL (Guillermo Leclair), RAS-NIC (Ilya Cardoza), AMUNIC (Roberto García López and Fernando de la Llana), ÉTICA Y TRANSPARENCIA (César Martínez and Luis Aragón), and USAID (Gerardo Berthin and Luz Marina García)—dedicated time and resources to this effort, thus contributing to the success of the program. Other donors such as CARE (Gerardo Martínez), CIDA (Maarten de Groot), SDC (María Luisa Pardo), and SIDA (Carlos Rivas) were also very supportive of this initiative in both countries, dedicating their time and expertise to the implementation of the selected projects. Ewen Le Borgne, from the International Water and Sanitation Centre (IRC) of Delft, the Netherlands, also contributed his time and support during the implementation of the program in Honduras.

Last, but not least, we would like to acknowledge the involvement of the peer reviewers, Dr. Patrik Stålgren of the Swedish Agency for Development Evaluation (SADEV) and Kathy Shordt, recently retired from the IRC, whose valuable comments were taken into account in this version of the report.

About the Authors

María González de Asís is a Senior Public Management Specialist on the Governance team in the World Bank Institute (WBI). Employed by the World Bank since 1997, she has concentrated on public sector reform. Most recently, she has managed anticorruption programs in the field, disseminating emerging best practice in governance and anticorruption worldwide, leading Legal and Judicial Reform Learning Programs, and researching and advising countries on governance and development. Dr. González de Asís has pioneered capacity-building approaches for good governance and anticorruption and is currently the governance and anticorruption coordinator for WBI. She is a frequent speaker on governance issues, having lectured at Stanford University, the Kennedy School at Harvard, and Georgetown University. Before joining the World Bank, she worked at Transparency International in Washington, DC; Berlin; and Peru; and for the Spanish law firm, Abogados Asociados, dealing with political anticorruption cases.

John Butterworth is a Senior Programme Officer at the International Water and Sanitation Centre in Delft, the Netherlands. He has worked in Latin America, Africa, and Asia on issues related to water and sanitation, water resources management, agriculture, natural resources management, and rural development. Previously Butterworth worked at the Natural Resources Institute of the University of Greenwich, U.K., as a water management specialist in the Livelihoods and Institutions Group, as a junior professional officer with the U.K. Department for International Development in Zimbabwe, as a PhD researcher (on catchment management in southern Zimbabwe) at the Centre for Ecology and Hydrology, and for Scott Wilson Resource Consultants as a hydrologist in the water engineering group. He has an academic background in soil water management and hydrology. Butterworth is a member of the Steering Committee of the Water Integrity Network and was a contributor to Transparency International's *Global Corruption Report 2008: Corruption in the Water Sector.*

Donal O'Leary joined Transparency International (TI) in July 2005 as a volunteer Senior Advisor, focusing on water sector issues. He was a founding member of the Water Integrity Network. While at TI, he has worked on multistakeholder anticorruption initiatives ranging from analyzing governance issues in large water supply and hydropower infrastructure in Africa to addressing corruption in municipal water systems in Central America. He is also one of the principal authors of TI's *Global Corruption Report 2008*. Before joining TI, he worked with the World Bank for more than two decades (1982–2005) on water and energy infrastructure projects in Africa and South Asia, including the preparation, implementation, and upgrading of hydroelectric and water supply projects in many countries. O'Leary holds engineering degrees from the National University of Ireland as well as a PhD in chemical engineering from the University of Texas (1971).

Dr. Per Ljung is Chairman and CEO of PM Global Infrastructure Inc. Before joining PM Global in 1998, he had a long and distinguished career at the World Bank, where he held a number of management positions. Trained in civil engineering, business administration, economics, and urban development, Ljung has spent most of his professional life working with infrastructure projects, with a major focus on the water sector. He has extensive experience with irrigation, river basin development, and rural and municipal water supply and sanitation. His book, *Energy Sector Reform: Strategies for Growth, Equity and Sustainability*, was published by the Swedish International Development Cooperation Agency in September 2007, and his article on "Water Corruption in Industrialized Countries: Not So Petty?" was featured in the *Global Corruption Report 2008*.

Abbreviations

ACODAL	Colombian Association of Environmental and Sanitary Engineers
ACP	Autoridad del Canal de Panama
ADB	Asian Development Bank
BPCB	Business Principles to Counteract Bribery
CONAPAS	Comisiôn Nacional de Agua Potable y Alcantarillado Sanitario (National Water and Sewerage Commission), Nicaragua
CONASA	Consejo Nacional de Agua Potable y Saneamiento (National Water and Sanitation Council), Honduras
ENACAL	Empresa Nicaragüense de Acueductos y Alcantarillado (National Water and Sewerage Company), Nicaragua
ERSAPS	Ente Regulador de los Servicios de Agua Potable y Saneamiento, Honduras
GAP	Gobierno Abierto y Participativo: Gobernando municipios sin corrupción program (Open and Participatory Government Program), World Bank Institute
GIS	Geographic Information Systems
GPS	Global Positioning System
IACAC	Inter-American Convention Against Corruption, OAS
IBNET	La Red Internacional de Comparaciones para Empresas de Agua y Saneamiento (International Benchmarking Network for Water and Sanitation Utilities)
IFEX	International Freedom of Expression Exchange
INAA	Instituto Nicaragüense de Acueductos y Alcantarillados (Nicaraguan Water Supply and Sewerage Institute)
MDG	Millennium Development Goals
NGOs	nongovernmental organizations
NRW	nonrevenue water
OAS	Organization of American States
OECD	Organisation for Co-operation and Development
OSCE	Organization for Security and Co-operation in Europe

PACTIV	Political leadership, Accountability, Capacity, Transparency, Implementation, and Voice
PEMAPS	Plan Estratégico de Modernización del Sector Agua Potable y Saneamiento (National Water and Sanitation Modernization Plan), Honduras
PET	public expenditure tracking
PPWSA	Phnom Penh Water Supply Authority, Cambodia
PROOF	Public Record of Operations and Finance
PUB	Singapore Public Utilities Board
SANAA	Servicio Autónomo Nacional de Acueductos y Alcantarillados (National Water and Sewerage Service), Honduras
TI	Transparency International
UN	United Nations
UNDP	United Nations Development Programme
WHO	World Health Organization
WSP	El Programa de Agua y Saneamiento (Water and Sanitation Program)
WSS	water supply and sanitation; water and sanitation sectors

The Nature of Corruption in the Water Sector

Per Ljung
Donal O'Leary
María González de Asís

Contents

MODULE 1

1 Introduction

1.1 Welcome

Clean water and adequate sanitation are essential for health and well-being. Unfortunately, poor governance of water and sanitation systems means that many peri-urban and rural areas lack services, and where they are available, the quality of service is unsatisfactory. The impacts of poor governance are many and easily identifiable. In many water systems, as much as half the water is unaccounted for—lost through illegal connections and excessive leakage. Tap water may be unsafe and may have to be boiled before it is fit to drink. Water may be available only a few hours a day, a few days a week. Many people in peri-urban and slum areas are forced to buy water from vendors at prices that are up to 10 to 20 times higher than prices for tap water. Rural villagers might have to walk for hours to collect water. Waterborne diseases kill millions of people each year.

The most serious governance problem is corruption, but corruption is not equally pervasive in all countries or water utilities. Similarly, the nature of corruption varies from one organization to another. Thus, the purpose of this course is to help participants identify corrupt practices and their underlying causes and prepare action programs to address the problem. Emphasis is placed on transparency in decision making, enhanced accountability for public officials, and improved information for the citizenry. Attention needs to be paid to these issues, independent of the reform model being used for the water sector, including privatization and decentralization.

1.2 Goals of the Module

The course is part of the World Bank Institute's Open and Participatory Government Program at the municipal level (*Gobierno Abierto y Participativo: Gobernando municipios sin corrupción,* or GAP).[1] It is aimed at people engaged in or interested in the water and sanitation sector in Honduras and Nicaragua. This is the first module of the course. It provides a conceptual framework for understanding the nature of corruption in the water sector and analyzing the effects of different types of corruption on customers, the sector itself, and the society at large. Several case studies illustrate both the basic concepts and the types of corruption that the participants can be expected to address in real life.

1.3 Learning Objectives

At the conclusion of the module, the course participants will be able to:

- Identify different types of corruption;
- Explain factors that tend to breed or encourage corruption;

- Analyze the nature of corruption in a water utility;
- Describe the impacts of corruption on the performance of the water company and, especially, on the urban poor; and
- Understand how corruption in the water sector can reduce growth and economic development.

1.4 Outline of the Module

To achieve these objectives, this module is organized as follows. Section 2 sets the context with a brief description of the water and sanitation sector in Honduras and Nicaragua. Section 3 discusses basic concepts such as "petty" and "grand" corruption and "individual" versus "systemic" corruption. This section also describes factors that cause corruption, such as distorted institutional incentives, lack of public information and transparency, and the lack of accountability.

To illustrate the concepts and the potentially pervasive nature of corruption, section 4 provides an in-depth case study of South India. It is augmented by a couple of briefer illustrations of ways to determine the extent and nature of corruption.

Section 5 presents a basic framework for analyzing corruption. This framework looks at the roles of different actors in the water and sanitation sector and examines the potential types of corruption in different activities from policy making and planning through construction to operation and maintenance, including billing and collection.

Section 6 analyzes the effects of different types of corruption in the short and long term. It also reviews the (limited) literature on the impact of corruption on service levels, costs, and economic performance.

Section 7 reviews the corruption and governance literature on decentralization and privatization and relates the results to the conceptual corruption framework presented in section 5.

Each section of the report includes a practical activity that gives the participant an opportunity to reflect on the material and examine how it relates to his or her own organization (or a water utility with which he or she is familiar). The module concludes with a more comprehensive activity that both integrates the various sections of the module and prepares the participant for subsequent modules.

1.5 Before We Start

Before starting, we would like you, the participant, to reflect on the situation in your own organization (or the water utility most familiar to you) by answering the following questions in Activity 1.1.

Activity 1.1

Some form of corruption exists in most water utilities. The nature and extent varies from one organization to another.

Question 1: Is corruption a problem in your organization (or in the water utility you are most familiar with)?

Question 2: What form does the corruption take? (For example, are construction or supply contracts awarded to the friends and family of the deciding official? Are meter readers "induced" to understate the level of water consumption? Are there illegal connections, and who has helped to create them? Are new projects selected because of political pressure rather than because of the needs of the population?)

Question 3: What are the consequences of the corrupt practices you have described?

2 Context

2.1 Introduction

Major progress has been achieved in expanding water and sanitation services to the urban and, to a lesser extent, rural areas in Honduras and Nicaragua. However, the sector is still experiencing governance problems that are reflected in poor service quality. To address these problems, the two governments have initiated reforms of sector institutions that are proceeding at varying speeds. To provide a context for the subsequent discussion of corruption, this section gives a brief overview of the emerging institutional structure in the two countries and highlights some of the more critical performance issues.[2]

2.2 Sector Organization

The sector structure in **Honduras** is still being transformed after the approval of a new Water Framework Law (*Ley Marco el Sector Agua Potable y Saneamiento*) in 2003. The implementation of the sector reform is guided by the National Water and Sanitation Modernization Plan (*Plan Estratégico de Modernización del Sector Agua Potable y Saneamiento,* or PEMAPS) prepared in 2005.

The National Water and Sewerage Service (*Servicio Autónomo Nacional de Acueductos y Alcantarillados,* or SANAA) traditionally operated about half the urban water and sewer systems, with the balance managed by municipalities. The 2003 law calls for the transfer of assets and the responsibility for operating more than 30 water and sewerage systems from SANAA to the municipalities by October 2008. SANAA's role will change from being an operator to becoming a technical assistance agency supporting the municipally owned utilities. Policy making will rest with the National Water and Sanitation Council (*Consejo Nacional de Agua Potable y Saneamiento,* or CONASA), and a new regulatory authority for water and sanitation (*Ente Regulador de los Servicios de Agua Potable y Saneamiento,* or ERSAPS) has been created.

All urban water systems are publicly operated, except in San Pedro Sula, where the city has granted a concession contract to a private company, and in Puerto Cortés, where the city has created a mixed (private-public) company. A couple of municipalities have also entered into lease arrangements with private operators. Under the decentralization scheme, it is anticipated that municipalities will establish autonomous operators for the water systems. PEMAPS stresses the need for strengthened regulation, improved governance, and transparency.

In principle, municipalities are responsible for providing water and sanitation services in **Nicaragua**. However, in practice only 26 smaller municipalities actually provide these services. Instead, most cities and towns are served by the national water and sewerage company (*Empresa Nicaragüense de Acueductos y Alcantarillados,* or ENACAL), which operates 147 separate systems. There are also three depart-mental water companies (in Jinotega, Matagalpa, and Rio Blanco) that together administer nearly 20 systems through management contracts with private compa-nies. Some 5,000 rural water supply systems are run by community organizations with support from FISE, the Emergency Social Investment Fund.

Under reforms initiated with new legislation in 1998, the National Water and Sewerage Commission (*Comisiôn Nacional de Agua Potable y Alcantarillado Sani-tario*, or CONAPAS) is in charge of policy making and sector planning, and the Nicaraguan Water Supply and Sewerage Institute (*Instituto Nicaragüense de Acueduc-tos y Alcantarillados*, or INAA) regulates the sector through concession agreements.

As noted above, there is a modest amount of private sector participation in water supply and sanitation. However, in 2003, the legislature severely limited the scope for the further involvement of the private sector. It suspended "the awarding of all concessions to private individuals for operation of ENACAL's facilities and assets, or the awarding of management contracts to private individuals" and it changed ENACAL's status from a "state-owned business" to a "state-owned public utility."

2.3   Access and Service Quality

Although both Honduras and Nicaragua have made major progress in extending water services to the urban population, rural water supply remains more problem-atic, as does sanitation in both urban and rural areas (table 1.1). While the service levels (official access) lag behind those of richer countries in the region, they com-pare well with service levels in Asia and Sub-Saharan Africa.

The quality of service, however, is quite poor. Water is rationed in most *Hon-duran* cities under SANAA management and, according to the World Bank, water is supplied two times a week or even less in the summer. In 2000, according to the World Health Organization (WHO-UNICEF 2007), 98 percent of water systems in Honduras provided water on an intermittent basis for an average duration of 6 hours per day. Drinking water was being disinfected in only 51 percent of urban water systems, and only 3 percent of the collected wastewater was being treated, which led to pollution.

Water supply in roughly half of the localities monitored by INAA in *Nicaragua* is not continuous, and the share is higher in the summer. Urban drinking water qual-ity generally meets WHO standards. It is also estimated that 42 percent of collected wastewater is treated (but few households are connected to a sewerage system).

2.4   Operating Efficiency

Nonrevenue water—or water that is not paid for—is estimated at 50 percent in Tegucigalpa, the Honduran capital, and 43 percent in San Pedro Sula.[3] In the early

Table 1.1: Share of Population with Access to Water and Sanitation Services in Latin America, 2004
(Percent unless otherwise indicated)

Country	Income 2004 GNI per capita (US$)	Water Urban Improved supply	Water Urban House connection	Water Rural Improved supply	Water Rural House connection	Sanitation Urban Improved facilities	Sanitation Urban House connection	Sanitation Rural Improved facilities	Sanitation Rural House connection
Haiti	400	52	24	56	3	57	0	13	0
Nicaragua	**830**	**90**	**84**	**63**	**27**	**56**	**22**	**34**	**0**
Bolivia	960	95	90	68	44	60	39	22	2
Honduras	**1,040**	**95**	**91**	**81**	**62**	**87**	**66**	**54**	**11**
Paraguay	1,140	99	82	68	25	94	16	61	0
Cuba	—	95	82	78	49	99	50	95	25
Colombia	2,020	99	96	71	51	96	90	54	20
Dominican Republic	2,100	97	92	91	62	81	65	73	27
Guatemala	2,190	99	89	92	65	90	68	82	17
Ecuador	2,210	97	82	89	45	94	62	82	16
El Salvador	2,320	94	81	70	38	77	63	39	2

1

Peru	2,360	89	82	65	39	74	67	32	7
Brazil	3,000	96	91	57	17	83	53	37	5
Jamaica	3,300	98	92	88	46	91	31	69	2
Argentina	3,580	98	83	80	45	92	48	83	5
Uruguay	3,900	100	97	100	84	100	81	99	42
Venezuela, R. B. de	4,030	85	84	70	61	71	61	48	14
Panama	4,210	99	96	79	72	89	58	51	1
Costa Rica	4,470	100	99	92	81	89	48	97	1
Chile	5,220	100	99	58	38	95	89	62	5
Mexico	6,790	100	96	87	72	91	80	41	16
Trinidad and Tobago	8,730	92	80	88	67	100	19	100	—

Sources: Income data from World Development Indicators 2006; Water and sanitation data from WHO-UNICEF Joint Monitoring Programme (http://www.wssinfo.org/).
Note: —. Data are not available.

2000s, the water system in Tegucigalpa had more than 9 employees per 1,000 connections, which is a high number.

In Nicaragua it is estimated that 18 percent of the connections are illegal and 56 percent of the water supplied goes unbilled. ENACAL has 6.5 employees per 1,000 customers, which is nearly double the number regarded as acceptable for a company of this type (that is, 3 or 4 employees per 1,000 customers).

2.5 Tariffs and Financial Performance

Water and sewerage tariffs in Honduras are low, especially in municipal systems, which indicates that tariff setting in municipalities is prone to "political capture."[4] Tariffs barely cover operation and maintenance costs and subsidies are generally poorly targeted.

Tariffs charged by ENACAL in Nicaragua are high in relation to incomes. Still, the company is in poor financial health due to the operating problems discussed above. This has a serious impact on both new investments and operation and maintenance.

2.6 Governance

The above analysis indicates that the water utilities in both Honduras and Nicaragua face some serious governance problems. No surveys have been carried out that directly measure and describe corruption in the water sector in the two countries. However, both the World Bank's *Governance Indicators 2006* (especially the variable measuring a country's "Control of Corruption") and Transparency International's *Corruption Perception Index 2006* indicate that public administration in general is affected by corruption in these countries. On both scores, Honduras and Nicaragua perform well below average for the surveyed countries (table 1.2).

Table 1.2: Indicators of Corruption in Honduras and Nicaragua

Country	World Bank		Transparency International	
	Index	Rank	Index	Rank
Best country	2.49	1	9.6	1
Nicaragua	−0.62	133	2.6	111
Honduras	−0.67	140	2.5	121
Worst country	−1.79	204	1.8	163

Source: World Bank: Control of Corruption (range from −2.50 to +2.50) from http://www.worldbank .org; Transparency International: Corruption Perception Index (range from 1 to 10) from http://www .transparency.org.

Table 1.3: Corruption in Public Utilities in Honduras

Responses from enterprises and consumers regarding key public utilities	Percentage of respondents receiving high quality of service	Percentage of respondents who were made to feel that bribes were necessary	Amount paid in unofficial payments, (lempira)	Percentage of respondents who did not make formal complaints because they believed it would not make a difference
Enterprises				
Phone installation	35	17	3,319	31
Electric connection	44	7	2,506	31
Water and sewerage	49	5	650	30
Consumers				
Phone installation	37	8	706	25
Electric connection	39	6	339	21
Water and sewerage	42	5	496	20

Source: World Bank Institute 2002.

A diagnostic survey of governance and anticorruption in Honduras undertaken by the World Bank Institute (2002)[5] indicates that corruption is common in public utilities, including those in the water and sanitation sector (table 1.3). The same survey also shows that corruption is common in public sector procurement. More than one-third of private sector firms interviewed believed that corruption was frequent in public procurement and estimated that the bribes were around 12 percent of the contract value.

3 What Is Corruption?

3.1 What Constitutes Corruption?

Corruption exists in all societies, to varying degrees and in different forms. Some practices that might be regarded as corrupt in one country might be legally and socially acceptable in another. Thus, there is no universally agreed upon definition for corruption. Indeed, attempts to develop such a definition invariably encounter legal and political issues. A good starting point is the definition used by the World Bank (1998, 19–20): *"Corruption is the abuse of public office for personal gain."*

However, corruption does not take place only in the public sector; it also occurs in nongovernmental organizations and private enterprises. Falsifying water meter readings, for example, is a corrupt practice whether it takes place in a private water company or in a public utility. Consequently, Transparency International uses a broader definition: *"Corruption is the abuse of entrusted power for private gain."*

It is common to make a distinction between "petty" corruption and "grand" corruption (see box 1.1 for description of common forms of corruption). *Petty corruption* typically involves small payments made to secure or expedite the performance of routine, legal, or necessary action such as getting a water connection or having a repair attended to expeditiously. Staff might also supplement their salaries by providing services "informally" (for example, by "selling" water to water vendors or tanker operators, or helping install illegal connections). Also common are small bribes for falsified meter readings. Transparency International Bangladesh's Baseline Survey on Corruption (1997) shows that 60 percent of urban households either paid money or exerted their influence one way or another to get water connections and to correct water bills (in addition to their legal payment). The report also says that nearly one-third of urban households had their water bills reduced through an arrangement with meter readers. While petty corruption might involve very small amounts, the frequency of such transactions means that the aggregate amounts can be very large.[6]

BOX 1.1 **Common Forms of Corruption**

Bribery. Probably the most common form of corruption, bribery is *the giving of some form of benefit to unduly influence some action or decision on the part of the recipient or beneficiary*. Bribery can be initiated by the person soliciting the bribe or by the person offering the bribe. The "benefit" may vary from money or other valuables to less tangible benefits such as inside information or employment.

(Continued)

1

BOX 1.1   Common Forms of Corruption (*Continued*)

Collusion. This is an arrangement between two or more parties designed to achieve an improper purpose, including improperly influencing the actions of another party. The most common form of collusion is when bidders agree among themselves on prices and "who should win." This may or may not involve paying bribes to government officials so that they will turn a blind eye to the practice.

Embezzlement and theft. This is the taking or conversion of money, property, or other valuables for personal benefit. It might involve diversion of public funds to one's own bank account or stealing equipment from the utility's warehouse.

Fraud. Fraud is the use of misleading information to induce someone to turn over money or property voluntarily, for example, by misrepresenting the number of people in need of a particular service. A private concessionaire might misrepresent the number of households connected to the sewerage system in order to obtain more favorable treatment from the regulator. A contractor might use substandard materials in construction (with or without paying a bribe to the supervising engineer).

Extortion. Extortion involves coercive incentives such as the use of threat of violence or the exposure of damaging information to induce cooperation. Office holders can be either the instigators or the victims of extortion. Extortion can also take the form of an official threatening to cut off water supply or refuse to certify measurements at a construction site.

Abuse of discretion. The abuse of an office for private gain, but without external inducement or extortion. Patterns of such abuses are usually associated with bureaucracies in which broad individual discretion is created or few oversights or accountability structures are present. Abuse of discretion can also be found in bureaucracies in which decision-making rules are so complex as to neutralize the effectiveness of the accountability structures that do exist. In a situation of water scarcity, abuse of discretion might involve giving preferential treatment to one neighborhood over another.

Favoritism, nepotism, and clientelism. In general, these involve abuse of discretion. However, in these specific cases, the act is governed not by the direct self-interest of the corrupt individual, but by some less tangible affiliation, such as advancing the interest of family (nepotism); a political party; or an ethnic, religious, or other group. These practices occur often in the hiring and promoting of staff. However, they can also take the form of building a new water system in "the minister's village."

Source: Adapted from Transparency International and UN-HABITAT (2004).

1

Grand corruption involves much larger amounts and is seldom as visible as petty corruption because both parties usually go to great lengths to conceal the transaction. Whereas petty corruption typically involves low-level utility staff, grand corruption tends to involve politicians, senior officials, and higher-level engineering staff. Grand corruption is most common in the award of large contracts for civil works, equipment, or concessions to operate major water systems. Davis (2004) in her survey of corruption in the water sector in South Asia reported that "the value of kickbacks paid was fairly consistent among the sites we investigated—between 6% and 11% of the contract value, on average."

Grand corruption does not exist only in construction. It is also frequent in the purchase of equipment and materials. Although many allegations have been made regarding corruption in the award of infrastructure concessions, few have been proven in court. One case, from Grenoble, France, shows that corruption is not limited to developing countries. In 1996, the city's mayor and an executive of the multinational water company were together convicted of accepting and paying bribes in the letting of concessions to run the city's water supply and sewerage services. In this sector, grand corruption can also be practiced by "water cartels" that manipulate service by imposing water tariffs on a large scale.

It is important to understand how widespread and well organized corruption is. Luis Moreno Ocampo has classified individuals into three types: the "demons," the "saints," and the "honest but sinful."[7] A key issue is how "saints" or "honest but sinful" people interact with the "demons" within an organization. The three types of individuals are illustrated in figure 1.1.

In many organizations, the great majority of the employees are either saints or honest but sinful. One or a few individuals might be devils and take the opportunity to collect bribes in, for example, the award of a contract. This situation, called *individual corruption*, is depicted in figure 1.2.

Individual corruption can be handled by strengthening surveillance and control systems leading to the identification and removal of the corrupt official(s). More

---

Figure 1.1   Demons, Saints, and the Honest But Sinful

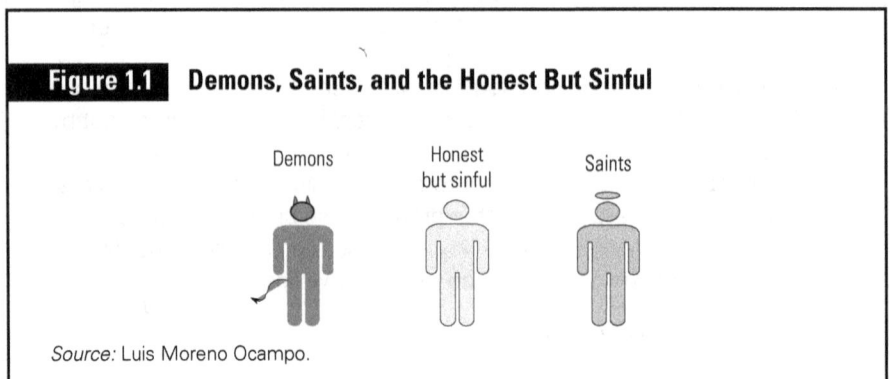

Source: Luis Moreno Ocampo.

serious is *systemic corruption* that permeates the whole institution. This occurs when the culture of the institution permits corrupt practices and typically involves not only staff and managers in a utility but also politicians and political parties. The spoils of corruption are often shared widely in the organization and with its political patrons. Systemic corruption is illustrated simply in figure 1.3. Note, however, that even when corruption is systemic, many employees are honest (or saints). There are usually also a number of employees who are not engaged in corrupt practices but who are well aware of the practices and, for a variety of reasons, do not report them or intervene in other ways.

Systemic corruption rarely exists in isolated cases. Rather, it tends to permeate the whole public sector, and norms and behavior are similar in most departments

Figure 1.2 Individual Corruption

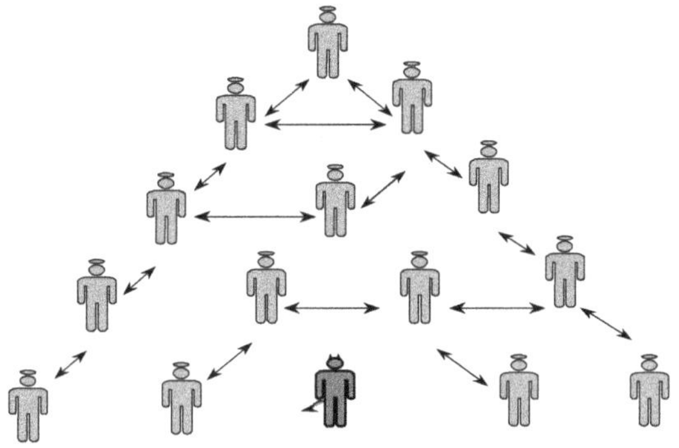

Source: Luis Moreno Ocampo.

Figure 1.3 Systemic Corruption

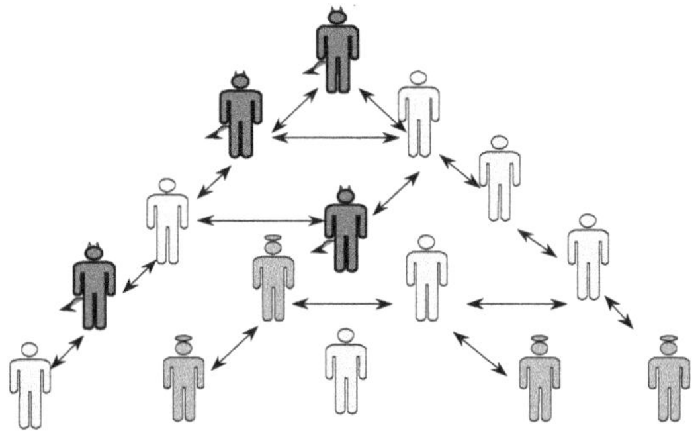

Source: Luis Moreno Ocampo.

and agencies. Bribery is known to be illegal; however, it is accepted as a usual practice in relations with the public sector. In systemic corruption, perverse incentives are very deeply rooted among public servants, and the resulting corruption is accepted by the private sector and the population at large as a way of doing business with government agencies.

The problem of systemic corruption cannot be tackled simply by removing corrupt individuals—the root causes must also be removed. Thus, traditional public sector management interventions need to be supplemented with transparency and related reforms as well as with wider engagement with the domestic private sector and civil society. Political will and leadership are essential for rooting out systemic corruption. Initiatives in the water and sanitation sector should be linked as much as possible to broader national anticorruption measures.

3.2 What Are the Root Causes of Corruption?

While it is commonly agreed that isolated cases of individual corruption are driven by greed, the phenomenon of systemic corruption is much more complex. Indeed, anthropologists, sociologists, political scientists, and economists typically describe the causes in different terms.[8] Klitgaard's corruption formula (box 1.2) captures in simple terms the key factors that make institutions susceptible to corruption.

BOX 1.2 Klitgaard's Corruption Formula

"Much can be said about the kinds of governments and, more generally, the kinds of institutions, be they public, private, or nonprofit, that are susceptible to corruption. Corruption tends to be reduced by the separation of powers; checks and balances; transparency; a good system of justice; and clearly defined roles, responsibilities, rules, and limits. Corruption tends not to thrive where there is a democratic culture, competition, and good systems of control, and where people (employees, clients, overseers) have rights to information and rights of redress. Corruption loves multiple and complex regulations with ample and uncheckable official discretion.

"Notice that most of these ideas apply to businesses as well as to governments. So does a metaphorical formula we find useful:

$$C = M + D - A$$

"**C**orruption equals **M**onopoly power plus **D**iscretion by officials minus **A**ccountability.

(Continued)

1

BOX 1.2 **Klitgaard's Corruption Formula (*Continued*)**

"If someone has monopoly power over a good or service and has the discretion to decide whether someone gets that good or service or how much a person receives, and there is no accountability whereby others can see what that person is deciding, then we will tend to find corruption. This is true whether we are in the public or the private sector, whether we are in a poor country or a rich one, whether we are in Beira, Berlin or Beirut."

Source: Klitgaard, MacLean-Abaroa, and Parris (2000).

Note: Klitgaard first proposed the formula in his book *Controlling Corrup-*

The water and sanitation sector is susceptible to corruption for the reasons outlined above. Uneven and inadequate coverage of services in most countries gives consumers an incentive to pay for a new legal or illegal connection. Furthermore, new construction projects tend to be infrequent, large, and unique, which means that little information is available upon which to judge how reasonable prices are. Projects are also complex, involving many subcontractors, and implemented in phases, which means that cost controls are difficult. Furthermore, in construction, information is asymmetric: the contractors and bidders know more about the real prices than the buyer. Further increasing the scope for corruption is the fact that most of the works are hidden underground.

Still, it appears that the water sector is no more susceptible to corruption than other government services in countries with systemic corruption. In terms of petty corruption affecting customers, water supply typically is ranked in the middle in corruption perception surveys. In terms of grand corruption in contracting and internal corruption in promotions and the like, the water and sanitation sector appears to be comparable to other types of public utilities and public works.

Activity 1.2

Looking at the water utility in your town (or another company that you are familiar with), please answer the following questions:

Question 1: Is corruption "individual" or "systemic"? Explain the reasons for the classification.

Question 2: Is "grand" or "petty" corruption the most serious problem? Explain why you believe this is the case.

Question 3: Which is the main cause of corruption: poor institutional incentives; lack of public information and transparency; or lack of accountability, tribal preferences, or constituency preferences? Explain.

4 Case Study: Corruption in the Water Sector in South Asia

4.1 Introduction

There have been very few in-depth studies of corruption in the water and sanitation sector. The most comprehensive examinations were undertaken by Jennifer Davis and presented in her article on "Corruption in Public Service Delivery: Experience from South Asia's Water and Sanitation Sector." This section builds on her article and quotes liberally from it.

Davis studied nine water organizations in South Asia (eight in India and one in Pakistan). Four were semiautonomous water boards that served large metropolitan areas and one was a municipal department. Four organizations served rural areas, one of which also served small towns and another of which served all urban areas in the state. Davis and her team conducted interviews with 350 staff members and 730 customers of these organizations. In addition, they had meetings with more than 320 elected officials, researchers, activists, journalists, and development professionals in the region. Regarding the methodology followed, Davis states:

> When questioning public agency staff and customers, interviewers used carefully designed and pre-tested questionnaires that placed sensitive questions toward the end of roughly one hour sessions. Respondents thus had an opportunity to develop rapport with interviewers before being asked about corruption related issues. At the same time, debriefing exercises with our interviewers suggest that a proportion of respondents were visibly uncomfortable answering questions about corruption and may have understated the occurrence of such behaviors. Rather than second-guess the veracity of respondents' answers, I present the raw data as collected, noting that these are likely conservative estimates of the incidence of such practices.

Davis focused on three types of corruption:

- Petty corruption associated with meter readings, repairs and, the expediting of new connections
- Bribery and kickbacks in contracting
- Internal corruption concerning the market for transfers

Thus, the survey does not cover all types of corruption that can be found in water companies. Furthermore, as Davis (2004) pointed out, the reported incidences of corruption (and the amounts involved) are likely to be conservative, that is, to present a lower limit for the frequency of corrupt transactions and the amounts involved.

1

4.2 Petty Corruption

The type of corruption reported most frequently were payments for falsifying meter readings. Some 41 percent of the respondents said that they had made at least one such payment during the preceding six months. Of the interviewed staff from the water agency, nearly three-quarters said that falsification of meter readings "happens about half the time," "is very common," or "happens virtually all the time." Partly reflecting the very low water tariffs in South Asia, the average amount paid was small—averaging about 45 cents per transaction.[9]

Some 30 percent of all consumers reported that they had made at least one payment in the previous month to expedite attention to repair work. Perhaps reflecting the value people put on water service (as opposed to what they pay for it), the median amount paid to accelerate repairs was $1.90. In one city, customers who paid no bribe to fix a problem with water leaks or backed-up sewage had to lodge a median number of four complaints before the problem was fixed. Virtually all of these complaints were lodged in person, requiring time during the workday and possibly the cost of transportation. For those making informal payments, the median number of complaints was two. Consequently, it is not surprising that many customers did not even bother to complain to the agency but handled the problem themselves or hired somebody to attend to it.

Some 12 percent of the respondents stated that they had paid bribes to expedite a new connection. The median payment for this was $22. The low frequency of payments for new connections might result from the "scheme" approach used for building distribution networks, in which a whole neighborhood is connected in one operation.

Davis (2004) makes the following comment regarding petty corruption and illegal connections.

Most managers in W&S [water and sanitation] agencies…expressed disapproval regarding bribery for meter reading and repairs among their staff. By contrast, very few considered the tolerance of illegal service connections to be a serious offense. In one city, we learned about a vigorous private plumbing market in which a household can obtain a water supply connection for roughly half the official fee charged by the public provider.…[However, as one] junior engineer explained, "A 'normal' illegal connection cannot be disconnected, so the customer is not afraid and will not pay."… Indeed, even when illegal connections are reported to the W&S agency, usually only a warning or a trivial fine is levied against the offender. Emphasis is placed on "regularizing" the household, i.e., convincing it to become a paying customer of the service provider. Staff of W&S agencies are under intense political pressure against disconnecting households with illegal connections, who are generally perceived to be lower-income residents.

During the interviews with staff from the water agencies, quite conflicting views were expressed. Davis reports:

> Among 80 respondents in one urban W&S agency, 52% agreed with the statement that, "Almost everyone uses contacts or money to get better services or special treatment," while 31% agreed that payments from customers to employees in exchange for faster service "benefit the customer and the employee without harming anyone else." Many pointed out that, in South Asian culture, the exchange of favors and small amounts of money, both in one's public and private life, is both commonplace and unobjectionable. As one field technician summarized, to staff at his level, petty corruption is generally viewed as "small potatoes." "No one calls this corruption, even. [Corruption] is happening at the higher levels" where greater opportunities for rent-seeking exist.

Others, however, expressed concern. One former director of an urban water board places great emphasis on addressing petty corruption:

> We must try to improve our public image.... [The] people must perceive us as honest. Otherwise, how can we make a case for increasing the tariff? The customer says "I am having to pay an extra 100 rupees just to have my repairs made on time." We cannot have this kind of image and expect public support.

A mid-career engineer in a state water department agreed:

> It is true that these payments do not involve a lot of money. But they involve the people's trust. It is corruption at their doorstep. It is corruption they can describe...first hand. So if you can reduce this kind of corruption they will know it. They don't have to take your word for it.

Despite the engineer's claim that these incidents of petty corruption do not involve a lot of money, box 1.3 shows that the in the aggregate petty corruption can involve much more money than grand corruption.

4.3   Corruption in Procurement

In every one of the studied institutions, some form of competitive bidding was employed. However, the process was generally flawed in at least one of three possible ways: bidders' cartels, political influence in contractor selection, or outright bribery in the bidding process.

BOX 1.3 Is Petty Corruption Really "Petty"? An Illustration from the Power Sector in Bangladesh

A recent survey of households in Bangladesh undertaken by Transparency International's Bangladesh Chapter (2005) found that

- 10.4 percent of households that have an electrical connection are using electricity with an illegal connection
- 70 percent of households that got an electrical connection in the last year had to pay 1,174 taka ($20) on average in bribes
- 4.3 percent of households that have an electrical connection paid an average amount of 1,445 taka ($25) to tamper with meter readings in order to avoid paying the amount due.

By 2002, Bangladesh had around 6.5 million households connected to the electricity grid, a number that was growing by 10–12 percent per year. Based on the survey data, the illicit payments for new connections amounted to around $10 million. The households bribing meter readers paid around $7 million. Presumably, the cost of an illegal connection was higher than the amount paid for "undermetering." If the annual cost for an illegal connection is assumed to be $30, the total amount collected by the linemen and meter readers would be around $20 million. Thus, in round numbers, we estimate that the Bangladeshi households in 2002 paid around $37 million in bribes. The payments made for the correction of billing errors, repairs of faults, and the like should be added to this number. While no data are available on bribe paying by industrial and commercial consumers, information from other countries in South Asia indicate that such entities bribe and steal more than households. Since households consume around 41 percent of the electricity billed in Bangladesh, it seems reasonable to assume that "petty" corruption in the Bangladesh power sector is a $100 million-a-year business.

In a report on governance in Bangladesh, the World Bank and UNDP (2002) quote a local businessman who reported that the prevailing bribe to obtain an equipment supply contract was 6–8 percent of the contract's value. For works contracts, the price was much higher—"as much as 20 percent." Assuming that 10 percent is a reasonable average for bribes involving the annual $300 million investment program in the electricity sector, it is likely that "grand" corruption amounts to around $30 million annually. (It should be noted, however, that TI-Bangladesh (2000) quotes higher figures: *"The rate of corruption is generally thought to be between 15 and 20% of the value of procurement contracts, although in limited cases it was reportedly as high as 30%."*)

In aggregate terms, probably some $130–150 million is collected each year in bribes in Bangladesh's electricity sector, with petty, rather than

(Continued)

1

| BOX 1.3 | **Is Petty Corruption Really "Petty"? An Illustration from the Power Sector in Bangladesh (*Continued*)** |

grand, corruption accounting for the greater share. However, this is only the tip of the iceberg. In 2002, there were more than 1,000 hours of planned load shedding and numerous hours of random black-outs. A survey of the investment climate in the country found that erratic and poor-quality electricity supply was the dominant constraint to business development and growth in Bangladesh (World Bank and BEI 2003). Another study carried out by USAID in 2002 valued lost production from power outages at US$772 million a year, equivalent to 1.7 percent of the gross domestic product

Davis (2004), quoting a contractor, describes the bidding cartels in the following words:

A group of [contractors] meet on the weekend in the office. We have a list of contracts being offered by [the public W&S agency]. We draw names out of a bag to see who will be the winner for each contract. That person decides what he will bid for the contract, and everyone else bids something higher than that.

The pre-determined winners of the contracts reimburse the losers for their bidding fees. The few contractors who were willing to provide such information estimated that the values of winning bids are roughly 15% higher than what would be bid in a competitive environment.

According to the contractors, bid rigging was most frequent in medium-sized contracts. Large contracts were "too important for anybody to forgo," and for small contracts, the potential gains were too small to make collusion worthwhile.

The water agencies adopted various strategies for overcoming this problem, such as using a "rate book" listing average and maximum unit costs for materials (such as pipe, cement, pumps) and works (trench digging, tank construction, and the like) and rejecting bids that exceeded the engineering estimate by a certain percentage (typically 20 percent). However, the success was limited. Another strategy was to split a certain job into many smaller contracts, but this approach often sacrifices economies of scale and may not result in any savings.

On larger contracts, where bid rigging is less frequent, contractors adopt different tactics. Davis describes it in the following terms:

[C]ontractors compete against one another by partnering with elected officials and senior bureaucrats, who can provide insider information and/or carefully manipulate tender documents to subvert even the best tendering systems.

1

The involvement by politicians in the process is described as follows:

Among the few contractors who admitted to making payments to politicians for assistance in winning tenders, the value of those payments ranged between 1% and 6% of the contract value. Others said that the quid pro quo can also take the form of non-cash exchanges. In one case, '[the official] wanted a water line extended to a colony that was not included in the project. We agreed that he would help us get the contract and we would do this extra work.'

Payments made to elected officials are recovered by contractors during contract execution, typically through the use of substandard materials and/or over-invoicing. Because the scale of these projects is large, detecting fraud is challenging even with reasonably good auditing procedures.

Although the technical staff of the water agencies tends to emphasize corruption that does not directly involve them, Davis (2004) also documents the kickback system involving the engineering staff.

Through complex arrangements funds budgeted for construction are "skimmed" and shared by a number of different actors. Contractors often pay either a percentage of the contract value or a lump-sum amount to one or more actors within the agency. The payments are almost always made in cash, in the W&S agency offices or in the field. Notably, in most agencies, staff reported that such payments are made only after contractors have been issued payment for completed work. Technical staff, thus, has a stake in seeing that construction works proceed apace, and the kickback system provides some impetus for timely completion of projects. Because these employees also have a vested interest in facilitating the processing of contractors' bills, they also often advocate with their agency's accounting departments to ensure prompt review and remittance of payments…The value of kickbacks paid was fairly consistent among the sites we investigated—between 6% and 11% of the contract value, on average.

The system of kickbacks is highly institutionalized, with well-accepted norms for who should receive what amount:

In one agency the schedule of payments for contracts valued up to US$44,400 is 1% of the contract value to each of six or seven staff members involved with the project (for a total payment of 6–7% of the contract value), starting with senior engineering staff and ending with the technical field supervisor. For contracts whose value exceeds US$44,400, lump-sum payments of between US$220 and US$1,100 are made to the same set of individuals (with senior staff receiving higher amounts). In virtually every W&S institution we visited contractors and agency employees confirmed that similar practices have occurred throughout their careers. One contractor produced a laminated card upon which he had written the payment schedule for kickbacks. "It is too hard to remember all the rules," he explained. "I don't want to make a mistake and pay any more than I have to."

Clearly, the contractors will recover the cost of various types of kickbacks through higher bid prices, overinvoicing for quantities produced, or using substandard materials. The "split" between these methods of "cost recovery" is not documented in the study. It should be noted, however, that the economic cost might differ significantly. The most expensive failure is probably when substandard materials and construction methods are used, leading to a shortening of the shelf life and increased operation and maintenance costs.

4.4   The Market for Job Postings

In part to reduce the scope for corruption, most public works agencies in South Asia have a policy of transferring staff every two or three years. Yet, this system has resulted in a thriving market for desirable posts. According to Davis (2004),

> staff…has developed a remarkably sophisticated calculus to estimate the value of a particular post (its extra-salary revenue generating potential) and thus the maximum amount they are willing to pay to secure a transfer.…Prices for different kinds of posts appear to be well established. In state-level agencies where the range of possible transfers is comparatively larger, a "plum" post (e.g., to a construction division within a desirable geographic location) costs the equivalent of four months' salary. The price of a position in construction or procurement located in a less desirable part of the state was 2.5 months' salary.…Payments are not simply made at the end of each two- to three-year posting. As one staff member explained: "If I want your position I can get help from someone to have you transferred out, even if you have been there less than two years. You will be told that someone wants your post and is willing to pay a certain amount for it. If you can pay more than that, you will keep your post."

The internal job market in the water agencies is, in many respects, integrated links in a complex web of systemic corruption, involving not only the bureaucracy but also politicians and other influential individuals.

> Very few staff reported paying their superiors for such transfers; instead, monies are given to politicians or unelected local leaders, who exert influence (and sometimes share part of the fee) with higher-level bureaucrats.…Of

course, cash is not the only currency with which staff can transact in the transfer market. One mid-level engineer described his authorizing water supply connections to a group of households on unregistered land (where public services are prohibited by law) in exchange for an assembly member's assistance with a transfer request. Another said that he provided several tankers of water without charge to a wedding celebration for a local leader, who in turn helped the staff member keep his post for a period beyond the typical three-year transfer threshold.

Activity 1.3

Looking at the water utility in your town (or another company that you are familiar with), please answer the following questions:

Question 1: How often do different types of bribe payments occur in your organization?

Question 2: How do the amounts paid in South Asia compare with those in your organizations?

Question 3: Are there some types of corruption that are not described in the case study? How serious are these types in your organization?

1

5 A Framework for Analyzing Corruption

5.1 Introduction

The foregoing sections have demonstrated that corruption in water and sanitation agencies can take many forms and involve many actors both within and outside the agency. Not all corruption involves bribes or illegal payments, however. It can also involve exchange of services and political favors. Not all forms of corruption impose the same cost on society; politically motivated "white elephants" might be more costly than illegal water connections, for example. The water sector is also complex with many players besides the government water utility. These players range from donors and central ministries to small-scale private providers and households with their own well.

Most government anticorruption interventions occur either at the national level (in the form of various anticorruption drives) or at the utility level. However, there is also a need to examine and address corruption at the sectoral level. Thus, this section provides a comprehensive framework for analyzing corruption in the water and sanitation sector. This framework, which has been developed by Plummer and Cross (2007), can be used to map corrupt practices in different settings by identifying who is involved, at what stage of service delivery the corrupt practices occur, and the links between various occurrences of corruption. Ultimately the goal of this sort of exercise is to provide a robust framework that is relevant and applicable to the sector, one that integrates project-level and crosscutting governance diagnostics and is usable as a tool for understanding and promoting change.

5.2 A Conceptual Framework

The corruption framework (illustrated in table 1.4) is structured around interactions and the type and level of activity, starting with policy formulation at the national level, then moving to project or program formulation and to the minutiae of billing and collection. The interactions involve all public, private, and civil society actors and institutions. The interactions of the public officer or agency are classified into three types in the framework:

- Other public actors or agencies
- Private actors or companies
- Consumers or civil society

In principle, if a private company owns or operates a water utility, there would be a fourth interaction—between the private company and its staff on one side and consumers and civil society on the other. However, the corruption that is likely to be

Table 1.4: A Framework for Analyzing Corruption

Venue	Public-Public	Public-Private	Public-Consumer
Policy making	• Policy capture (competition and monopolies)	• Policy capture	
Regulation		• Regulatory capture (such as waivers of regulations and licensing)	
Planning and budgeting	• Distortions in decision making by politicians (affecting location and types of project investments) • Corruption in national and sector planning and budget management (misuse of funds, interministerial bribery for fund allocation, collusion or bribery in selection and project approval) • Corruption in local budget management (fraud, falsification of accounts or documents, village level collusion)	• Bribery to influence allocation of resources • Bribery in sector budgeting management (influencing, distortions in funding allocation), national and local • Bribery to delay debt restructuring	
Donor financing	• Donor-government collusion in negotiations to meet spending and funding targets • Donor-government collusion and fraud with regard to progress and quality	• Donor and national private operator collusion (outside legal trade agreements)	
Fiscal trans-fers	• Bribery, rent seeking, and kickbacks to ensure fund transfers between ministry of finance and sector ministries		
Management and program design	• Corruption in personnel management: payments for preferred candidates (e.g., utility directorships), payments for promotions and transfers, salary perks	• Collusion between agency staff and consultants to bias the result of design and cost studies or environmental and social assessments	• Influence project decision making • Bribery for preferential treatment, elite capture

(Continued)

29

Table 1.4: A Framework for Analyzing Corruption (*Continued*)

Venue	Public-Public	Public-Private	Public-Consumer
	• Distortionary decision making (collusion with leaders in selection and approval of plans/schemes) • Corruption in local government and departmental planning and budget management	• Bribery to delay efficiency operations	• Distortionary decision making at the project level (such as site selection, equipment, construction)
Tendering and procurement	• Administrative corruption (fraud, falsification of documents, silence payments) • Interdepartment or agency collusion over procurement and construction	• Bribery to influence contract or bid system • Corruption in award of concessions and in decisions over duration, exclusivity, tariffs, subsidies • Corruption in procurement: inflated estimates for capital works, supply of chemicals, vehicles, equipment • Falsification of documentation	
Construction		• Not building to specification • Failure to complete work • Underpayment of workers • Fraudulent invoicing	• Corruption in community-based construction (with similar types of practices as for public-private interactions)
Operation and maintenance		• Overbilling by suppliers, theft or diversion of inputs • Avoiding compliance with regulations, specifications, and health and safety rules • Falsification of accounts	• Installing or concealing illegal connections, avoiding disconnection, receiving illicit supply • Administrative corruption for speed (or preferential treatment) of repairs/new connections.
Payment (for services)			• Fraudulent meter reading • Overcharging

Source: Authors.

found in this case is of the same nature as that where a public water company and its staff interact with private contractors, suppliers, and consumers and civil society.

5.3 Public-to-Public Interactions

Corruption involving interactions between public servants or public institutions can take many forms. At the higher levels of government, this type of corruption is generally opaque and complex. The "personal gains" do not have to be monetary. They can also be in the form of "intangible" benefits such as status, job security, and political influence. During the Suharto and Marcos regimes in Indonesia and the Philippines, respectively, the whole government machinery cooperated in adopting policies, programs, and projects that directly or indirectly benefited the family and friends of the presidents. These were clear cases of "state capture," where a small clique of wealthy and influential people benefited greatly from government actions. While private actors obviously offered some inducements, much of the public-public corruption involved pressure and interference from politicians and senior bureaucrats on the lower, executing levels. Corruption in Peru during the Fujimori regime appears to have operated in a similar manner.

Various forms of public-to-public corrupt practices can be found in policy-making functions in many countries. Politicians and officials responsible for water sector policies may seek to influence the focus of policy and investment priorities to set up future opportunities for rent seeking or election gains. Large dams might be favored over cheaper groundwater development for water supply, for example.

Regulators can be influenced by politicians and other stakeholders to set standards and regulations that benefit the utility at the expense of the consumer and the public (regulatory capture) or to allow projects to bypass established standards or procedures. Officials are expected to "play the game," and their status and power base is dependent on their willingness to work within the established system.

Engineering staff may manipulate budgets—with the connivance of the financial staff—to increase the expected income from maintenance contracts.

Diversion of public funds through outright theft and through practices such as "ghost workers" can be found wherever internal controls are lax, purposely or otherwise. Some equipment and material is easily stolen for personal use or sale.

Bribes for promotions, appointments, transfers, and a multitude of perks are common in many bureaucracies throughout the world. Buying senior appointments is thought to be frequent throughout Sub-Saharan Africa, and the prices paid for utility directorships or municipal engineers are often common knowledge and calculable, based on sector norms. Many argue that these types of practices, common throughout the civil service, lie at the core of the incentive and patronage system and propagate other forms of corruption. Corrupt politicians and managers might also appoint willing personnel to lucrative positions on the condition they pass on a portion of their "side" income.

31

Another example of public-public corruption occurs when a foreign government intervenes (through its embassy) to obtain favorable treatment of its firms in getting regulatory approvals and award of infrastructure concessions and consultancy or construction contracts.

5.4 Public-to-Private Interactions

Procurement requires interaction between the public and private sectors and is the most publicized face of corruption. Every level of government and every type of government agency has to purchase goods and services, normally from the private sector. In the water and sanitation sector, a number of public actors may be involved depending on the size and type of project: national and local government politicians and managers; municipal engineers, operations staff, project managers and procurement officers; and a set of private actors that might include suppliers, contractors, operators, and local and national consultants.

While public attention typically focuses on the award of contracts, corruption in procurement and project implementation is a much more complex phenomenon. It occurs throughout the process from initial design to issuance of completion certificates. A partial list of corruption opportunities in procurement includes:

- Design specifications (especially for equipment) might be deliberately biased so that only one supplier can meet the specifications (only one manufacturer, for example, might be able make a machine with the specified number of reverse gears).
- The method of procurement can be manipulated in favor of sole sourcing (common in consulting) or through "change orders" to increase the scope of jobs rather than new bidding.
- Like design specification, prequalification requirements can be biased in favor of one or a small group of bidders.
- The method for advertising the contract and time allowed for bidding might limit competition (for example, by making it difficult for foreign firms to bid).
- Bid evaluation criteria might be biased in favor of one firm ("in-country" experience, for example).
- Collusion among bidders is commonly done with the full understanding of public officials (who expect their own "rewards" during project implementation).
- Bidders can obtain inside information that helps them to produce bids that are more likely to win or to arrange a bidding cartel.
- Bid evaluations are easily manipulated, especially in the case of consultancy contracts. However, bid evaluations for construction and equipment and materials supply contracts can also be manipulated. Bidders can, for example, be disqualified on flimsy (but hard-to-refute) evidence of "noncompliance." Contractors can falsify records and documentation to ensure bids

look competitive, and officials may either encourage them to do so or turn a blind eye. This practice is facilitated by the secrecy that usually surrounds bids and bid evaluation.

- Even where public procurement rules forbid direct negotiations concerning the price, there are often direct contacts with the bidders for "clarifications" or negotiations of technical and contractual terms.
- Various types of cheating take place during project implementation—with the connivance of water agency staff. Quantities might be overstated; specifications are not followed, and so on.
- Change orders can be issued for both legitimate and illegitimate reasons, typically enhancing the profits of the contractor.
- Most construction firms in the world are thinly capitalized and their cash flow is of utmost importance. This situation creates great opportunities for extortion by engineering staff who can refuse to certify progress or issue completion certificates. Financial staff can exert pressure by delaying progress payments.

The system of kickbacks can be very elaborate. One approach is to require foreign consulting firms to employ a local partner (to "build local capacity"). The local firms do very little work and grossly overcharge their foreign partners. Much of that revenue earned is then paid to the local officials. Although the foreign firms might be seen as "victims" in this arrangement, in practice, it allows them to circumvent any legislation in their home countries that might prevent them from directly paying bribes.

Operation and maintenance involves procurement of repair works, material, and equipment, as well as services such as billing and collection, security, and cleaning. The interactions with contractors and suppliers providing these goods and services fall prey to the same types of corrupt practices as seen in the procurement and construction of new works.

Private ownership or operation of water and sanitation systems does not mean that the opportunities for corruption disappear. Corrupt deals have occasionally been negotiated behind closed doors. Even when concessions and service contracts are awarded through "competitive" bidding, the scope for corrupt practices is large and similar to that experienced in the procurement of large works. Once a concession or private management arrangement is in place, the private operator may falsify operating records to earn higher fees or make greater profits. Even if no bribes or kickbacks are provided, the private operator(s) can unduly influence the regulatory authority to provide favorable rulings (a situation commonly referred to as regulatory capture).

Small private providers in large and small towns and peri-urban areas present an opportunity for an alternative set of corrupt interactions. These interactions occur in the water market between public (local government and utility) officials and small private providers of water (such as water truckers and traditional water vendors). These operators usually depend on the water utility for their supply, which creates the possibilities for irregular activities in the access to and payment for water. In some cities, the water dealers have created virtual cartels that limit competition and fix prices—at the expense of the urban poor.

In rural water supply, powerful individuals can "expropriate" the local well (with the "approval" of local officials) and sell water to the villagers at exorbitant prices. In the sanitation sector, small-scale private operators may pay local government officials to allow them to dump waste on inappropriate sites irrespective of health and environmental consequences.

5.5   Public-to-Consumer and Civil-Society Interactions

The corrupt interactions that take place between public water sector officials and water consumers are petty, frequent, and systemic (they may be either extortive or collaborative).[10] The main types (falsifying meter readings, expediting repairs, and installing new connections as well as illegal connections) were extensively covered in the South Asian case study in section 4 and will not be repeated here. It should be noted, however, that kickbacks for falsified meter readings are not the only problem associated with billing. In some cities, consumers consistently report problems with overbilling, requiring them to go to the local utility office (and pay a bribe) to get the bill corrected.

Status and political influence also has a major impact on the quality of water supply. It is not uncommon to find, for example, that high-income neighborhoods have a continuous 24-hour water supply and a functioning sewerage system while slum areas might get water (through stand pipes) only for a few hours a couple of times a week and completely lack any sanitation facilities.

In rural areas, corruption can affect the design, implementation, and ongoing maintenance of water supply and sanitation projects supported by the community and nongovernmental organizations. Political influence rather than a needs-based analysis may determine the location of new schemes. Village leaders may collude with government overseers to obtain preferred access and placement of the water point at a location convenient to the elite. Although a community organization or village committee is typically expected to operate such schemes, these organizations are frequently captured by a local strongman who manages the affairs of the system for private gains. Community-based systems in urban areas suffer from similar patterns of behavior, distorting the type of installation selected and its management.

Activity 1.4

Please complete the "corruption matrix" below for the water agency you are most familiar with, answering Yes or No. For cases where you answer Yes, you may wish to provide more details.

Venue	Public-public	Public-private	Public-consumer
Policy making			
Regulation			
Planning and budgeting			
Donor financing			
Fiscal transfers			
Management and program design			
Tendering and procurement			
Construction			
Operation and maintenance			
Payment (for services)			

1

6 Impact of Corruption

In general terms, studies have shown that systemic corruption tends to slow down overall economic growth, reduce local and foreign investments, and increase income inequalities.

The most obvious effect of corruption in the water and sanitation sector is that it increases the cost of service delivery. Construction costs are increased and so are operation and maintenance costs. Davis (2004) estimated that water agencies in South Asia might spend 20–35 percent more on construction contracts than the value of the services rendered. Estache and Kouassai (2002) estimated that nearly two-thirds of the operating costs for 21 water companies in Africa were attributable to corruption. Corruption also influences project selection and slows down implementation, further reducing the direct economic benefits and the financial viability of water utilities. Poor project selection can be very costly. For example, a study in Malawi by Water Aid found that if the current financial resources were effectively targeted at the unserved, usually poor areas, the Millennium Development Goal (MDG) target for water service could be achieved even if the level of investment fell 30 percent below that seen during the last five years. If the work was not targeted, reaching the MDG target for water would be both more expensive and possibly unachievable (Stoupy and Sugden 2003).

Leautier, Kaufmann, and others (2006) examined the performance of infrastructure services (water, sewerage, electricity, and telephones) in 412 cities in 134 countries. The authors find that corruption has significant and substantial effects on both access to services and on the quantity and quality of service delivery. Increased costs and reduced revenues from water utilities mean that less money is available for extending the service.

Unfailingly, in cities and towns, it is the poor who feel the greatest impact, typically by having to buy water from vendors and tankers at prices up to 10–20 times higher than the amount the better-off pay for tap water. Studies have also shown that the poor tend to pay a greater share of their incomes in bribes than the rich.

A study by Kaufmann, Montoriol-Garriga, and Recanatinil (2005) found that corruption hit the poor the hardest in Peru. They concluded:

> The evidence presented suggested that corruption may act as a regressive tax and that quality of governance is linked to access to public services. In particular, we constructed new measures of governance using data from users of public services from 13 government agencies in Peru. We found that for certain basic services low income users pay a larger share of their income than wealthier ones; i.e. the bribery tax is regressive. …[I]n the case of basic services, low income users appear to be discouraged more often and not to seek such a basic service than wealthier ones. Bribery may penalize poorer users twice over, first by acting as a regressive tax, and then as a discriminating

mechanism for access to basic services…. The public agency-level analysis suggested that corruption reduces the supply of services, while voice mechanisms and clarity of the public agency's mission increases it. [11]

Corruption increases the cost of providing water and sanitation services. Because aid and government budgets are limited, fewer people can be served as a result. To make up the shortfall, as much as an additional $48 billion might be needed to meet the Millennium Development Goals for water supply and sanitation (Plummer 2008). Furthermore, poor maintenance of pipes and intermittent supply increase the risk of contamination and waterborne diseases for those already connected. The social costs of inadequate water supply and sanitation in developing countries are extremely high. According to the United Nations Development Programme (UNDP 2006), some 1.8 million children die each year from diarrhea, and together unclean water and poor sanitation are the world's second biggest killer of children. Remote, poor rural regions of a country are the areas that tend to be underserved when it comes to village water supply, affecting women and girls who may have to walk for hours to collect water.

1

Activity 1.5

Looking at the water utility in your town (or another company that you are familiar with), please respond to the following questions:

Question 1: To what extent do various corrupt practices affect the performance of the water and sanitation services? Try to link certain practices to the cost, quality, and reliability of service.

Question 2: Are certain groups affected more than others? Can you describe these groups and explain why you believe this is the case?

7 Sector Restructuring and Corruption

Governance and corruption are intrinsically intertwined. Poor governance breeds corruption and corruption erodes attempts to improve governance. Indeed, it might be argued that in countries and institutions with systemic corruption, the governance system is designed to create opportunities for corruption. Whatever the case, the nexus of poor governance and corruption in the water and sanitation sector limits access and leads to a deterioration in the quality and reliability of services. It also reduces people's faith in public institutions and undermines the legitimacy of the government. The wider social costs by far exceed the gains of corrupt practices.

To improve the performance of water and sanitation utilities, many governments have undertaken major restructuring of the sector. Two approaches that directly address governance structures and institutional incentives have been adopted. One is to split up large national agencies and decentralize construction, operation, and maintenance to locally managed companies. The other is to bring in the private sector through management contracts, concessions arrangements, or a complete divestiture. Each approach has its proponents.

Advocates of decentralization stress the benefits of shortened lines of communications, quicker decision making, and greater accountability to the local beneficiaries. Decentralization, the proponents argue, means that those who are hardest hit by corruption can take part in the decision-making process, thus removing some of the incentives to engage in corrupt practices. Moreover, decentralization should increase the level of information available for management and oversight, and a closer relationship between service providers and their clients can increase the moral cost of corruption. However, as the skeptics argue, this argument disregards the role that locals sometimes play in the creation of corruption. Decentralization can also lead to a proliferation of public offices and with it the number of officials who can exercise their powers for private gain. Moreover, evidence suggests that the close interactions between public officials and consumers created by decentralization entails a personalization of relationships with a high degree of patron-client characteristics that may enable corrupt behavior.

The relationship between decentralization, good governance, and corruption, is in part an empirical question. What factors are most important in a given setting? Shah (2006) examines both the conceptual and empirical basis of corruption and governance. He also uses the word *localization* as a more precise term than *decentralization*. Localization implies home rule, that is, decision making and accountability for local services at the local level. Shah concluded that:

> ...decentralized local governance is conducive to reduced corruption in the long run. This is because localization helps to break the monopoly of power at the national level by bringing decision-making closer to people. Localization strengthens government accountability to citizens by involving citizens in

1

monitoring government performance and demanding corrective actions. Localization as a means to making government responsive and accountable to people can help reduce corruption and improve service delivery. Efforts to improve service delivery usually force the authorities to address corruption and its causes. However, one must pay attention to the institutional environment and the risk of local capture by elites. In the institutional environments typical of some developing countries, when in a geographical area, feudal or industrial interests dominate and institutions of participation and accountability are weak or ineffective and political interference in local affairs is rampant, localization may increase opportunities for corruption. This suggests a pecking order of anti-corruption policies and programs where the rule of law and citizen empowerment should be the first priority in any reform efforts. Localization in the absence of the rule of law may not prove to be a potent remedy for combating corruption.

The proponents of privatization of water, sewerage, and other infrastructure services argue that the greater technical and management skills found in the private sector, combined with a profit motive, will lead to greater efficiency, better quality of services, and less corruption. Indeed, some studies show that private water utilities are more efficient and provide better service than public utilities. However, the evidence from cross-country data is inconclusive. It is clear that the performance depends on a large number of factors, including the form of private participation and the regulatory arrangements. Using enterprise-level data on bribes paid to utilities in 21 transition economies in Eastern Europe and Central Asia, Clarke and Xu (2002) found that bribes paid to utilities were higher in countries with greater constraints on utility capacity and lower levels of competition in the utility sector and where utilities were state owned. In other words, privatization reduced corruption.

This finding is not very surprising, since it could be expected that the profit motive would induce managers to clamp down on employees who receive bribes for corrupt practices such as falsifying meter readings. It should be noted, however, that the contractual arrangements play an important role in providing incentives for managers to do so. Under management and *affermage* (leasing) contracts, the state remains the owner of the assets, is responsible for new investments, and receives a share of the revenues. In such cases, the private operator has virtually no incentive to keep construction costs down and a smaller incentive to reduce corruption by the meter readers than under concession arrangements.

Opponents of water privatization, however, argue that the award of concessions and other types of private participation contracts can be as corrupt as the award of large construction contracts. There is also the risk of regulatory capture, where the regulatory agency, for one reason or another, tends to protect the interests of the private operator over those of the state and consumers.

Thus, using the interaction framework presented in section 4, we find that decentralization and privatization break some of the existing corruption links. At the same time, however, new—potentially corrupt—links are created. This highlights the need to carefully design any decentralization and privatization processes to achieve a significant improvement in the governance of the water and sanitation sector.

Irrespective of the basic institutional arrangements (centralized, decentralized, or private), the last decade has demonstrated that reforms to strengthen user rights, enhance stakeholder participation, and increase transparency in utility budgeting and decision making and can significantly improve utility performance and reduce corruption. These measures are discussed in module 3.

1

1

Activity 1.6

Considering the general political and institutional conditions in your country and the arrangements for governance of the water and sanitation sector, please answer the following questions:

Question 1: Can you propose measures to reduce the risk of corruption in local, decentralized water and sanitation enterprises?

Question 2: Can you propose measures to reduce the risk of corruption in private water and sanitation enterprises?

8 Concluding Activity

We hope that you have found the material in this module useful and that it has helped deepen your understanding of corruption. To give you an opportunity to once again think through the material, we suggest you complete the final activity below.

Activity 1.7

1

Looking at the water utility in your town (or another company that you are familiar with), please answer the following questions:

Question 1: What are the most serious types of corruption in terms of their impact on consumers, especially the urban poor?

Question 2: Can you propose suitable actions to address each of these problems?

1

Notes

1. For an overview of GAP, see http://info.worldbank.org/etools/docs/library/208546/GAPOverview.pdf.

2. This section is based primarily on documents from the World Bank (2007) and Inter-American Development Bank (2006) augmented by the entries on water supply and sanitation in Honduras and Nicaragua from Wikipedia (http://en.wikipedia.org/wiki/Water_supply_and_sanitation_in_Honduras and http://en.wikipedia.org/wiki/Water_supply_and_sanitation_in_Nicaragua).

3. Nonrevenue water (NRW) comprises three components: physical (or real) losses, commercial (or apparent) losses, and unbilled authorized consumption. The World Bank database on water utility performance (IBNET, the International Benchmarking Network for Water and Sanitation Utilities, at www.ib-net.org) includes data from more than 900 utilities in 44 developing countries. The average figure for NRW levels in developing countries' utilities covered by IBNET is around 35 percent (Kingdom, Liemberger, and Marin 2006). In a well-managed water system, the NRW would normally be below 20 percent.

4. Under political capture, regulation becomes a tool of self-interest within government or the ruling elite.

5. Also available in Spanish. See World Bank Institute (2002).

6. See box 1.3 in section 4 below for an illustration of how petty corruption can add up to "grand" amounts.

7. Ocampo is the former head of Transparency International's activities in Latin America and the Caribbean.

8. See Andvig and Fjeldstad (2000) and Stålgren (2006) for overviews of the different perspectives on corruption and its causes.

9. The average urban water tariff covers less than one-third of the operations and maintenance costs. In New Delhi, for example, the tariff is around 4 cents per cubic meter. (Unless otherwise indicated, all dollar amounts are U.S. dollars.)

10. Most of the public-to-consumers corrupt practices described here can also occur when the utility is privately owned and managed.

11. The authors also found that the water agency ranked number 2 among the 13 agencies in discouraging low-income customers from seeking service.

MODULE 2

Diagnosing Corruption in the Water Sector: Tools and Impact Indicators

Per Ljung
Donal O'Leary
María González de Asís

Contents

MODULE 2

1   Introduction

1.1   Welcome

This module focuses on diagnostic tools and impact indicators that help give a general picture of governance performance within an entity rather than focus simply on the nature and extent of corruption. Many of the factors that breed corruption, such as distorted institutional incentives, lack of accountability, and inadequate public information and transparency, result from poor governance and weak management systems. Thus, improvements in overall governance can not only reduce corruption but improve the performance of water utilities to the benefit of society at large. A serious study of the nature and extent of corruption and related governance problems can in itself have a direct impact by increasing political and management commitment to eliminating corrupt practices and improving the performance of the utility.

2

Until now, little systematic work has been done on corruption in the water sector. However, the nature of corruption in electricity and irrigation sectors and in local governments is very similar. Thus, the tools for water utilities presented in this module are largely adapted from other sectors.[1]

1.2   Goals of the Module

This module will provide the reader with a set of diagnostic tools and impact indicators that can help identify corruption and other governance problems in the water sector.

1.3   Learning Objectives

At the conclusion of the module, the reader will be able to:

- Identify and analyze the incentives and mechanisms that promote corruption
- Analyze the water utility's decision-making system
- Study governance and institutional performance patterns
- Identify the areas of the water utility that are most vulnerable to corruption
- Work with external diagnostic instruments
- Work with internal diagnostic instruments
- Propose measures to strengthen the role of regulatory agencies in improving sector governance and combating corruption.

1.4   Outline of the Module

Section 2 provides an introduction to the topic of identifying corrupt practices, assessing the vulnerability of the utility to corruption, and evaluating the impact of poor governance and corruption.

Section 3 presents a number of external diagnostic tools—tools used to obtain information from actual and potential customers as well as from other groups such as suppliers, contractors, and prospective employees—primarily aimed at capturing the nature and extent of corruption and its effects.

Section 4 presents internal tools that seek to assess institutional factors that enable corruption, such as leadership, institutional culture and incentives, accountability, internal procedures, regulation and enforcement, public information, and transparency of operations.

Section 5 discusses the important role that regulatory authorities (and other organizations with responsibility for sector oversight) play in collecting and publishing data on utility performance.

Each section of the module includes a practical activity that gives the participant an opportunity to reflect on the material and examine how it relates to his or her own organization (or a water utility with which he or she is familiar). The module concludes with a more comprehensive activity that both integrates the various sections of the module and prepares the participant for subsequent modules.

1.5 Before We Start

Before starting, we would like you, the participant, to reflect on the situation in your own organization (or the water utility most familiar to you), by answering the questions in activity 2.1.

2

Activity 2.1

Some form of corruption exists in most water utilities. The nature and extent varies from one organization to another. Corruption can also result from, as well as cause, more general problems of poor governance.

Question 1: What are the signs that corruption exists in your organization (or in the water utility you are most familiar with)?

Question 2: If you were to be asked to help reduce corruption in your organization (or in the water utility you are most familiar with), what kind of information would you like to have?

2

Question 3: Is it possible to address the causes and effects of corruption in this organization without addressing a more general problem of poor governance?

Question 4: How would you monitor progress in combating corruption?

2 Overview of Tools and Instruments

2.1 Introduction

The first step in addressing corruption is a diagnosis of the scale and nature of the problem. As could be seen in module 1, corruption can potentially involve purely internal processes, as well as interactions with politicians, other government entities, customers, suppliers, and contractors. It is essential to identify not only areas where corrupt practices actually occur but also those potential vulnerabilities where, if not counteracted, corruption might occur in the future.

Corruption is essentially illegal, which means that it is performed "in the dark" or "in the shadows," where it rarely can be observed directly. With the exception of customers, most of the bribe givers and bribe takers are unlikely to speak freely about the problem. Thus, any diagnosis of corruption involves making inferences based on incomplete data.

The first of several **warning signals**, or red flags, in a water utility is a high level of **unaccounted-for water**, or the losses in the system.[2] The losses broadly fall into two categories. Physical losses from leaky and broken pipes result mainly from technical inefficiencies or inadequate finances to replace aged water pipes. Commercial losses are mainly caused by water theft through illegal connections and falsified meter readings that in turn are the results of poor management and corruption. A system in reasonably good shape should have physical losses in the range of 10–20 percent and negligible commercial losses. Losses on the order of 50 percent, as found in many systems in Honduras and Nicaragua, indicate a major managerial problem stemming at least in part from false meter readings and unauthorized connections. While a high level of unaccounted-for water is a warning signal, a low level of reported losses does not necessarily mean that there is no problem with corruption.[3]

Other performance measures, such as the **number of staff per 1,000 connections**, can be useful indicators of governance problems, especially if they are used to compare similar utilities in a country. Indeed, most of the indicators used in formal benchmarking of water and sanitation utilities (see section 4.4) can be viewed as red flags. Newspaper articles, letters to the editor, and complaints from community groups can also warn the utility's management of serious governance problems.

While fighting corruption might not be part of its formal mandate, **the regulatory authority** can play an important role by increasing transparency and creating an operating environment that promotes good governance. One way this can be achieved is to collect, verify, and publish a consistent set of performance indicators for all water utilities. This topic is further discussed in section 5.

2.2 A Quick Analysis

Depending on the time and the resources available, an agency can decide to undertake either a quick analysis or a detailed one.

In many cases, a **quick analysis** will be sufficient to begin to address corruption and governance problems in a water utility. A quick analysis is most appropriate in smaller and medium-size utilities where financial resources are limited. The analysis is based on the premise that the water agency's staff and its customers have a fairly good understanding of the quality of service and potential governance problems. The starting point typically is one or two participatory workshops, similar to focus meetings, with representatives from civil society and utility staff. To be productive, these workshops should have a well-defined thematic agenda based on the objectives of the diagnosis, but with enough flexibility to follow up new leads. The workshops should be led by an experienced and neutral facilitator. The workshops with civil society focus on the utility as a service provider, that is, on the quality of service, the responsiveness of staff, and the suitability of procedures as well as on perceptions of corruption. The objective of the workshop(s) with utility staff is to examine the problems of governance and rendering of services, thus identifying perverse incentives and prioritizing the areas most vulnerable to corruption. On the basis of this information, the employees may themselves recommend the necessary institutional reforms and corrective actions.

In the quick analysis, the participatory diagnosis can be complemented by a review of internal documents such as procedural manuals, budget papers, and procurement documents to identify areas susceptible to corrupt practices and reforms that can reduce corruption risks.

2.3   Detailed Analysis

A **detailed analysis** consists of a set of instruments that have been developed using stringent techniques to obtain the most exhaustive and objective description possible. Besides offering reliable information for formulating reform measures to guide the intervention, this empirical evidence also contributes to depersonalizing and depoliticizing the approach to corruption as a problem in the public institution. The relative disadvantage of a detailed analysis is the higher cost and greater time needed for its implementation.

2.4   Classification of Diagnostic Tools

As we have seen earlier, there are many types of corrupt practices that involve a broad range of actors. The various diagnostic tools described in the subsequent sections tend to capture different types of corruption. Based on their main focus, the diagnostic tools tend to be classified as either external or internal. **External diagnostic tools** seek to collect information from utility's customers and suppliers as well as from civil society and the business community in general. This can oftentimes be difficult given the water sector's separation from the wider public and their civil society organizations. **Internal diagnostic tools** are concerned primarily with the organization's own policies, procedures, and incentives that present opportunities for and occurrences of corruption. The key resource is the organization's own staff, as well as written documents.

Efforts to improve governance and reduce corruption can have different focuses and objectives, depending in part on who is driving the effort. That in turn will have an impact on the design of the diagnostic tools used. If, for example, the chief executive is trying to improve the performance of the water utility, the focus will be on this single utility. If a mayor of a city is seeking to upgrade city services, water and sanitation might be included in a multisectoral diagnostic analysis. The drive for improved governance and reduced corruption might also be spearheaded by the central ministry in charge of the sector or by the regulator. In this case, the diagnosis would be limited to a single sector but might cover a number of water and sanitation companies located in different cities and towns. Indeed, water sector legislation, policies, and procedures have a major influence on governance and corruption throughout a country, and thus central institutions have an important role to play.

2

Activity 2.2

Looking at the water utility in your town (or another company that you are familiar with), please answer the following questions.

Question 1: If asked to examine corruption in this organization, would you undertake a quick analysis or a detailed diagnostic study? Please explain your reasons.

Question 2: Where and how would you find most of the essential information on the nature and extent of corruption in the organization? By examining internal documents? Talking to managers and staff? Interviewing customers and/or suppliers and contractors? Can this information be gathered in group meetings, or is it best gathered through confidential interviews? Please explain your reasons.

Question 3: Who should take the lead in addressing corruption and governance problems in the organization? Please explain.

3 External Diagnostic Tools

3.1 Introduction

External diagnostic instruments are those used to obtain information from actual and potential customers as well as from other groups such as suppliers, contractors, and prospective employees. Organizations that speak for these groups—nongovernmental organizations (NGOs), community groups, business and trade associations, and the like—might also be covered. The idea is to gain an understanding of the client's perceptions of the water utility's performance in various areas. The principal external instruments are:

- Corruption surveys
- Citizen report cards
- Participatory corruption assessments

In practice, the distinction between different diagnostic tools is fluid. Depending on the nature and extent of governance and corruption problems, a survey instrument can be tailored to address a multitude of issues. Formal surveys also can be complemented with focus group discussions and a limited number of free-form, open-ended interviews.

The choice and design of instruments depend largely on the purpose of the assessment (to mobilize public opinion, for example, or to form the basis for concrete action plans or to compare different utilities) and the available budget. It will invariably involve trade-offs and compromises.

3.2 The Corruption Survey

The corruption survey is a tool that helps identify unethical practices by highlighting ordinary people's perceptions of corruption. It can be applied at different levels: a specific water utility; all services in a city (water supply, electricity, health care, and so on); all water and sanitation systems in a country; or a broad range of sectors on a nationwide basis. This assessment is essential not only for formulating strategies that address existing problems but also for developing systems that ensure greater transparency in the future.

A corruption survey has the following key objectives:

- To identify the organizations, institutions, or sections within institutions where corruption is prevalent
- To quantify the costs of corruption to the average citizen
- To increase public interest in the issues surrounding corruption

2

- To provide a basis for actions to be taken in the light of the findings of the survey
- To provide an objective yardstick against which progress can be measured.

The corruption survey can be designed to capture the perceptions of corruption in the institution(s) of interest among the public at large (actual and potential customers), enterprises that use the services, employees of the institution and other civil servants, and its suppliers and contractors. Quite naturally, the survey instrument (questionnaire) needs to be tailored to each target group.

The results of the survey(s) need to be interpreted with care. Households and enterprises are likely to be quite open about payment of "speed money" for a new connection or repairs and of "tea money" to get a billing error corrected. They might be less forthcoming about more clearly illegal practices like bribes for falsified meter readings or unauthorized connections. Staff and enterprises engaged in grand corruption have reasons not to answer questions regarding their own behavior. Thus, questions addressed to these groups are usually expressed in generalized terms like "in your organization" or "contractors" rather than in direct personal terms. Even then, when corruption is systemic, the responses to such questions tend to understate the frequency of such occurrences and the amounts involved. Table 2.1 summarizes the kind of issues addressed in a corruption survey.

The questionnaire should be carefully designed and pretested to solicit as accurate results as possible. The selection of interview subjects should follow sampling procedures commonly used in social science research.

Table 2.1: Issues Covered in a Corruption Survey

Issue	Description
Frequency of interaction	Relevant organizations that the respondent has interacted with in the past year and how often (once a month or more often, less than once a month, or only once in the past year)
Purpose of interaction	The purpose of interaction could be classified into categories along the following lines: a. Obtain a new connection b. Repair/service problems c. Paying bill d. Correcting bill e. Employment f. Other
Bribery incidence	Whether bribes are required or demanded to obtain or expedite services (or avoid law enforcement) and what the respondents expect the consequences to be of declining to bribe (that is, satisfactory service, bad service, harassment, delay, or denial of service)

(Continued)

Table 2.1: Issues Covered in a Corruption Survey (*Continued*)

Issue	Description
Bribery transaction	The actual bribes that the respondents have paid or know others (friends, business associates, or competitors) to have paid. Respondents provide information on the amount, the frequency (every day, at least once a week, at least once a month, at least once in the past year), and the purpose as classified above
Corruption trend	Relevant organizations in which respondents have perceived improvement or deterioration in the level of corruption, the magnitude (small, moderate, or large), and the period over which the change has been perceived (last year, last three years, last five years)

Source: Authors.

2

It is important to present a comprehensive report on the responses to various questions. However, if one purpose of the survey is to identify the most or least corrupt sector in a city or country or the most or least corrupt water utility in a country, it is often appropriate to combine various corruption measures for a specific service entity or sector into an overall "corruption index."[4] Any single index will involve subjective judgments of the relative importance, or weights, of its respective components. (Even an "unweighted" summary of different components implies a judgment that all components are equally important.)

For an illustration of a sectorwide survey, see the case study on corruption in the water sector in South Asia, which was presented in section 4 of Module 1.[5]

3.3 The Citizen Report Card

Often corruption is only one of many governance problems afflicting a water utility. These problems can include limited access to water, poor quality of the water, unreliable supplies, a high amount of unaccounted-for water, many billing errors, slow response times to service problems, long waiting lists for new connections, and poor esteem for the utility staff. In these cases, it is often appropriate to take a broader approach than a corruption survey and to cover all aspects of the water service.

Various participatory systems for analyzing public services by citizens have emerged over the years. One of these is the citizen report card, which was pioneered in the Indian city of Bangalore (box 2.1). It has been applied successfully in a number of countries, including in Honduras (*Sistema de Indicadores para el monitoreo de los derechos económicos, sociales y culturales en los servicios públicos de agua potable y energía eléctrica*), prepared by El Centro de Investigación y Promoción de los Derechos Humanos, a nongovernmental organization based in the country.

Citizen report cards are instruments to encourage public accountability. Modeled on a private sector practice of conducting client satisfaction surveys,

---

BOX 2.1 **The Use of Citizen Report Cards in Bangalore, India**

Frustrated by the poor quality of public services, a group of private citizens in Bangalore, India, decided in 1994 to undertake a survey to collect feedback from users of central and local government services in the city. The services ranged from the police and government hospitals to the telephone, electricity, and water utilities. The success of this initial effort led to the creation of the Public Affairs Centre, which subsequently developed the methodology for the citizen report card and helped spread its use throughout the world. In Bangalore, the survey was repeated in 1999 and 2003. The progress between the first and second surveys was relatively modest.

However, *The Third Citizen Report Card on Public Services in Bangalore* (2003) documented a striking improvement in the quality of service across the board. The percentage of people "satisfied" with the water and sanitation service increased from 4 percent in 1994 to 73 percent in 2003. Over the same time, satisfaction with staff behavior in the water sector rose from 26 percent to 92 percent. Between 1999 and 2003, the accuracy of water bills increased from 32 percent to 90 percent. For all city services, the percentage of people reported to have paid "speed money" fell from 23 percent in 1999 to 11 percent in 2003.

Several special factors existed in Bangalore that drove this improvement in services. The chief minister was committed to improving services, but the citizen report card gave him and city managers a tool to measure and monitor performance. The city had (and still has) a booming economy led by the information technology industry. This industry also attracted young, well-educated, and ambitious individuals whose norms and values were more like the ones found in Silicon Valley than those of the traditional Indian middle class. Thus, they were more likely than most people to voice their concerns or "vote with their feet" (by immigrating or settling in another Indian city with better services).

2

report cards solicit user perceptions on the quality, efficiency, and adequacy of the various public services that are funded by taxpayers. Qualitative user opinions are aggregated to create a "score card" that rates the performance of service providers. The findings present a quantitative measure of overall satisfaction and perceived levels of corruption among an array of other indicators. By systematically gathering and disseminating public feedback, report cards can serve as a "surrogate for competition" for monopolies—usually government-owned—that lack the incentive to be as responsive to their client's needs as private enterprises. They are a useful medium through which citizens can credibly and collectively "signal" to agencies about their performance and pressure for change.

2

Like the corruption survey, the citizen report card can cover a single water utility, all public services in a city, all utilities in the sector, or all public services throughout a country. The first option is appropriate if the manager of a water and sanitation utility seeks to upgrade the performance of the company. The second option is adopted if the mayor or the municipal council want to ensure that the city's citizens receive "value for their tax dollars." The third option can be arranged by, for example, the central policy-making entity (such as the ministry of water) or the regulatory agency. The fourth option is undertaken on behalf of the cabinet or the legislature to improve the overall performance of the government.

The citizen report card is intended to examine the services provided by surveying the recipients or beneficiaries of these services and rating the responses according to a scale that measures efficiency and value. The larger purpose of the report card tool is to use the results of the survey to advocate for improvements in the services provided and to further investigate the reasons behind the provision of inadequate services. By repeating the exercise every couple of years, the progress of various managers and entities can be monitored and compared.

In its simplest form, the survey conducted for the report card consists of a number of questions regarding the level of customer satisfaction with various aspects of the service, such as water pressure, water quality, and accuracy of billing. However, to provide a better basis for action by the utility's management and staff, the survey should also include some more objective measures of service quality and of its effect on the direct and indirect costs imposed on the customers. In connection with billing procedures, for example, the respondents could be asked questions along the following lines: Was your last bill correct? If not, did you report the problem? Why or why not? How did you report the problem? If you went to the water utility's office, how many times did you have to go? How far away is the office? How did you get there? How long did you have to wait each time? Was the problem finally resolved? Was the staff helpful? Was the staff courteous? Did you have to pay "tea money" to get somebody to correct the problem? Did you get help from a friend or relative to persuade the utility staff to correct the problem?

A report card survey needs to be carefully designed and executed and its results widely disseminated to both utility officials and the general public. Thus, such a survey will involve the following steps:

- Identify issues through focus group discussions
- Design the instruments and test them
- Identify the scientific sample for the survey
- Hire an independent (and credible) agency to conduct the survey
- Collect and analyze the data
- Place the results in the public domain
- Advocate and establish partnerships

3.4 Participatory Corruption Appraisal

Formal surveys, such as those described in this section, are excellent tools for collecting information that can easily be quantified. However, they are less suitable for capturing information of a more qualitative nature, such as how corruption (or poor service quality) affects the poor and how the poor cope with and respond to corrupt practices affecting their everyday lives.

Section 2.2 described how participatory workshops can be used for a quick assessment of corruption. The participatory corruption appraisal incorporates some of the same elements but carries them further. This tool focuses on the impact of corruption on the most vulnerable—the poor. It was first introduced in Indonesia (box 2.2) as part of a World Bank–supported initiative.

The general objectives of the participatory corruption appraisal are:

- To understand the harmful effects of corruption on the lives of poor people

2

BOX 2.2   Participatory Corruption Appraisal in Indonesia

In 2000–01, the Partnership for Governance Reform, a joint program of Indonesia and the World Bank, organized and developed an action research project called Corruption and the Poor. The project was undertaken in three urban slums in Makassar, Yogyakarta, and Jakarta and aimed to use participatory corruption assessment techniques to explore how corruption affects the urban poor in Indonesia.

In each location the project team talked to groups of 30–40 poor men and women about their experience with corruption. These group sessions were followed by individual interviews throughout the community to elicit where and how corruption affected them. These talks allowed immediate insight into poor people's lives and provided a holistic understanding of the direct costs of corruption. It also motivated the participants, as well as researchers, to stay engaged beyond the fieldwork, linking research to action.

The participants identified four major costs of corruption:

- Financial costs: Corruption eats into already tight budgets and therefore puts a higher burden on the poor than on the rich.
- Human capital: Corruption erodes access to and the effectiveness of social services including schools, health care services, food subsidy schemes, and garbage collection, to the detriment of poor people's physical well-being and their skills.

(Continued)

2

BOX 2.2   Participatory Corruption Appraisal in Indonesia (*Continued*)

- Moral decay: Corruption erodes the rule of law and reinforces a "culture of corruption."
- Loss of social capital: Corruption destroys trust and damages relationships, corroding community cohesion.

The activity took place in two phases:

1. The research phase consisted of a first visit to the communities for fieldwork and a second visit that linked research to action, where findings were reported back to each community and a process for follow-up action was kicked off.
2. The action phase consisted of several, location-specific follow-up activities, involving local nongovernmental organizations, media, and the community.

The Corruption and the Poor project resulted in two publications. The first, *The Poor Speak Out,* is a set of 17 journalistic pieces that record the stories of poor people who participated in the action research. The book outlines the types of corruption they confronted in their everyday lives and records how the individuals concerned chose to handle it. The book also includes an analysis of the results of the project. A second publication entitled "Participatory Corruption Appraisal" records the methodology used to engage poor urban communities and elicit from them what they thought were the most corrupt practices they encountered, and what they thought could be done about it.

A third aspect of the project was the involvement of local community organizations who wanted to work on anticorruption activities as a follow-up to the action research project. Civil society groups in Makassar and Yogyakarta spearheaded the establishment of networks of over 40 nongovernmental organizations, as well as universities and professional organizations, to take local action against corruption and help urban communities fight against corruption in the lower levels of the public service. Activities included working with popular theatre, community-based education, the mass media, comic strips, and alternative media channels to disseminate anticorruption messages and establishing corruption and public policy monitoring groups at the community level.

Source: TI and UN-HABITAT 2004.

- To communicate such information widely to policy makers and the general public
- To help the communities in which the appraisal took place to plan and act to reduce corruption

To carry out a participatory corruption assessment, it is important to collaborate with an organization that has the confidence of the poor in the general area in which information is sought. This will usually be an NGO that has an existing program in the area or a local community organization. This organization does not have to be active in a field related to water supply and sanitation. The important thing is that the local people trust it.

The team of field workers that would conduct the assessment is usually made up of individuals from the NGO or community association who are able and willing to learn about the methodology and carry it out in the community. This involves developing skills in participatory focus group discussions and interviews.

A number of focus group discussions are held over a period of one to two weeks. The group discussions are typically complemented by a number of in-depth personal interviews. The information collected must be sifted and organized in a way that can be easily understood.

The assembled information is then presented to the community, and possible actions are discussed. This could be followed up by a public meeting (with the community's agreement) in which the findings from the appraisal and the action plans proposed are presented to a larger audience (local government officials, local NGOs, traditional local leaders, local journalists). The idea is to amplify the voice of the community and seek others' involvement in addressing the problems caused by corruption.

2

Activity 2.3

Looking at the water utility in your town (or another company that you are familiar with), please answer the following questions.

Question 1: What are the strengths and weaknesses of the corruption survey?

Question 2: What are the strengths and weaknesses of the citizen report card?

Question 3: What are the strengths and weaknesses of the participatory corruption assessment?

Question 4: Which methodology would you use for a detailed diagnosis of corruption and governance problems in your organization? Please explain your choice.

4 Internal Diagnostic Tools

4.1 Introduction

The internal diagnostic and monitoring tools discussed is this section would normally be used in conjunction with one or several of the external tools discussed in the previous section. The following tools are presented:

- A utility checklist
- Vulnerability assessment
- Performance benchmarking
- PROOF: Public Record of Operation and Finance

2

The first two are quite simple and are most appropriate when the institution is experiencing a limited amount of individual corruption rather than pervasive systemic corruption. To be effective, they require not only that the water utility's management be firmly committed to rooting out corruption but also that the commitment be well known in the organization.

The last two tools are most appropriate when the water utility experiences broader governance problems and when it is essential to systematically improve service levels and to make the utility more responsive to the needs of the citizens in its service area.

4.2 The Utility Checklist

To minimize loopholes and opportunities for corruption within a water supply and sanitation company, it is useful to examine the realities and specific conditions that may sustain corrupt activities. The utility checklist focuses specifically on the utility's management system and aims to assess the vulnerability of the system to the abuse of authority and resources.

The purposes of the utility checklist are:

- To identify and begin to focus on the specific areas of vulnerability to abuse of authority and management of resources that a utility might have.
- To provide a common ground of information and understanding for all parties interested in knowing about and improving the effectiveness of the water utility. The dissemination of this information helps to promote transparency.

The utility checklist generates a profile of useful information obtained through (among other means) direct interaction with municipal officials and employees. Information that may indicate potential loopholes for corrupt practices should be widely disseminated to interested parties and the public at large.

The impact of decisions made and actions taken as a result of the checklist is intended to be helpful in achieving transparency at the utility level. Moreover, the process of obtaining information itself, as well as its wide dissemination, also plays an important role in encouraging transparency.

The checklist is a series of questions divided into sections that correspond to those areas of utility operations that have generally been most subject to abuse or in most need of strengthening to overcome corruption. They could include:

- The corporate ethical framework
- Public complaints mechanisms
- Leadership
- Human resources
- Service levels and targets
- Budgeting
- Procurement
- Audit procedures

Box 2.3 provides a sample of the questions, divided into the eight areas listed above, that may serve as a framework for creating questionnaires tailored to specific situations.

BOX 2.3   **Examples of Questions Covered in a Utility Checklist**

Ethical Framework
   1. Is there a code of conduct for the senior managers?
   2. Is it used and thought to be effective?
   3. Are the assets and incomes of senior managers disclosed annually to the public through effective means?

Public Complaints
   4. Is there an independent complaints office within the utility?
   5. Is it known to the public and to staff?
   6. Is it effective and respected?
   7. Is there retaliation against whistle-blowers, or are they protected?
   8. Can anonymous complaints be made?
   9. Is there a program for testing the integrity of the various departments or business units?
   10. Is the program publicized and is it effective?

Leadership
   11. Is the senior leadership committed to the fight against corruption and how has this been demonstrated in both words and deeds?
   12. Does the public respect the work of the utility?

(Continued)

BOX 2.3 **Examples of Questions Covered in a Utility Checklist**
(*Continued*)

Human Resources
13. Is there respect for work rules by all staff, including supervisors?
14. Is the system for recruiting, disciplining, and promoting staff fair?
15. Are pay scales and benefits fair?
16. Is the internal administrative system for appeals of staff decisions considered fair?

Service Levels and Targets
17. Are service levels in different areas monitored on a regular basis?
18. Are targets for service improvements set on an annual basis in consultation with the affected public?
19. Are actual service levels and service targets made public?
20. Are budget allocations clearly linked with service targets?

Budgeting
21. Is the budgeting process well publicized and open to the public?
22. Does the public actively and directly participate in shaping the utility's budget priorities?

Procurement
23. Is the procurement system reputed to be fair?
24. Is it based on competitive principles?
25. Are procurements advertised in advance and made known to the public?
26. Is the process for selecting a bidder thorough and fair?
27. Are conflict-of-interest rules enforced?
28. Does the utility make its investments through a competitive process?
29. Are certain types of procurements excluded from competition?
30. Have there been corruption issues with the procurement system?
31. Is there a regular audit of procurement actions?

Audit Procedures
32. Are the accounts regularly audited by independent auditors?
33. Is there an internal auditor?
34. Are the results made public in a timely and effective manner?
35. Is there a separate government public accounts committee supervising or reviewing the audit process?
36. As a result of these audits, are actions taken to rectify systems and practices?

2

It is quite important that a wide range of stakeholders, including relevant people within the municipal government and the regulatory authority, be asked to complete the utility checklist. There are many reasons for this, including the need for an accurate appreciation of the wider community's views on corruption within the organization, as well as the need to lay the basis for outside monitoring of the municipality's performance.

An assessment making use of the checklist can be carried out in a number of different ways. These can include small group meetings with individual work units, larger workshops, and outside studies. The underlying principle, however, is that the assessment must be conducted in a collaborative manner with the staff. Through a partnership approach, and not simply as an audit-type inquiry by an external consultant, the checklist can become a learning tool and an instrument for promoting change within the organization. An initial assessment will need to be made by the senior leadership, both political and civil service, to determine the approaches that will ensure maximum candor in responses.

The checklist can play an important role in helping various stakeholders understand the strengths and weaknesses of the water company's integrity systems. The ultimate goal is to have the results of the checklist serve as a basis for change within the organization. This can happen with certainty only when the leadership is committed to good governance and has in place the systems that will enable it to act effectively.

While the checklist is primarily a self-assessment tool, it can also serve the very important purpose of building an informed community. For this reason, it is recommended that stakeholder involvement be built into the process of conducting the checklist study. This could be done through interviews and focus groups, as well as through broader public meetings. Furthermore, sharing the results of the assessment based on the checklist and the steps to be taken by the municipality can go a long way in building trust between the stakeholders and local government and in enhancing transparency.

4.3 Vulnerability Assessment

The vulnerability assessment is another tool that a water supply and sanitation company and outside organizations can use to help them understand how the utility addresses integrity and transparency issues. This tool focuses on three areas: whether the general control environment is permissive of corruption; whether a particular activity is more likely to be susceptible to corruption; and whether existing controls are adequate.

The main purposes of the vulnerability assessment are:

- To clarify the different areas within the organization that might be vulnerable to the abuse of authority and management of resources
- To point authorities and reformers in relevant directions concerning the steps to be taken to reduce vulnerability, enhance transparency, and strengthen integrity

The vulnerability assessment generates information that is especially useful in identification of loopholes in the local system that allow corruption to occur in the utility. Thorough analysis can point to systemic changes for reducing corruption and enhancing transparency in the local government.

Similar to the utility checklist, the vulnerability assessment poses a series of questions, to be answered after thoroughly examining the utility structure. These responses are then analyzed to identify the areas of vulnerability. Finally, remedies are proposed to improve the general municipal environment and reduce risks of corruption in the pinpointed areas. Box 2.4 puts forth an outline for a vulnerability assessment under the three-pronged framework described above.

BOX 2.4   Examples of Questions Covered in a Vulnerability Assessment

A. Is the general control environment permissive of corruption?

- To what degree is management committed to a strong system of internal control?
- Are appropriate reporting relationships in place among the organizational units?
- To what degree is the organization staffed by people of competence and integrity?
- Is authority properly delegated and limited?
- Are policies and procedures clear to employees?
- Are budgeting and reporting procedures well specified and effectively implemented?
- Are financial and management controls, including the use of computers, well established and safeguarded?

B. To what extent does the activity carry the inherent risk of corruption?

- To what extent is the program vague or complex in its aims; heavily involved with third-party beneficiaries; dealing in cash; or in the business of applications, licenses, permits, and certificates?
- What is the size of the budget? (The bigger the budget, the greater the possible loss.)
- How large is the financial impact outside the agency? (The greater the "rents," the greater the incentives for corruption.)
- Is the program new? Is it working under a tight time constraint or immediate expiration date? (If so, corruption is more likely.)
- Is the level of centralization appropriate for the activity?
- Is there evidence of previous illicit activities here?

C. After preliminary evaluation, to what extent do existing safe-

---

2

4.4 Performance Benchmarking

Poor governance of a water supply and sanitation utility leads to excessive costs, lower revenues, and poor quality of service. The key problem for both utility managers and key stakeholders (for example, the municipal government, the regulatory authority, and the general public) is knowing whether the entity is well managed. For a "normal" private enterprise, the starting point would be to see if the company made a profit or not. However, water and sanitation services produce significant social benefits that are not captured in a simple profit figure. The next logical step would be to look at the cost of providing the service, but such a figure in and of itself does not say if the utility is well run. In the case of water supply, costs vary dramatically from one place to another: desalination might be required, for example, or the nearest fresh water source might be hundreds of kilometers away. The typography of the city might be very hilly, requiring extensive and costly wastewater pumping. To provide a meaningful measure of the performance of water supply and sanitation utilities, the process of "benchmarking" utilities has been developed. Benchmarking uses a broad range of performance indicators, defined in a consistent manner to allow comparisons with other entities that have similar characteristics and with industry leaders.

Benchmarking allows managers to identify and prioritize key areas for improvement, searching for best operating practices in these areas and adapting these practices through measures that improve one's own performance. Benchmarking is not a one-time exercise but rather a tool for continuous performance improvement that yields benefits when done systematically over a period of time. Moreover, the person conducting the benchmarking is important, as is what is done with the information gathered. Oftentimes benchmarking information is gathered, but it is unclear to the public what should be done given what has been learned. Information gathered should serve as a basis for some sort of improvement. Benchmarking is now widely used across the public and private sectors for a variety of objectives, including efficiency improvements in systems and processes, optimizing costs, organizational restructuring, among others, ultimately enhancing the quality of services or outputs that are delivered to the customer.

There are two approaches to benchmarking: metric and process. Metric benchmarking, which is the most common approach, is a quantitative comparative assessment using standard performance indicators. (For an illustration of metric benchmarks, see box 2.5 and box 2.6.) Process benchmarking involves identifying specific work procedures to be improved through step-by-step "process mapping" and then locating external examples of excellence for standard setting and possible emulation. Metric benchmarking identifies the performance gaps and desired levels to be attained, whereas process benchmarking

2

BOX 2.5 **Benchmarking: Service and Performance Indicators**

Service Coverage
Water coverage
- Total population with access
- Population with household connection
- Population served by public water points

Sewerage coverage
- Population with household connection

Production and Consumption
Water production
- Liters per person per day
- Cubic meters per household per month

Water consumption
- Total as liters per person per day
- Total as cubic meters per household per month
- Residential/household consumption as percentage of total
- Industrial and commercial consumption as percentage of total
- Institutional and public uses as percentage of total
- Bulk supply as percent of total
- Total residential/household consumption as liters per person per day
- Residential consumption for household connections as liters per person per day
- Residential consumption for public water points as liters per person per day

Nonrevenue Water
- Unaccounted-for water expressed as a percentage of net water supplied
- Volume of water "lost" per kilometer of water distribution network per day
- Volume of water "lost" per water connection per day

Metering Practices
- Proportion of connections that are metered
- Proportion of water sold that is metered

Network Performance
- Pipe breaks per kilometer per year
- Sewerage blockages per kilometer per year

(Continued)

2

2

BOX 2.5 **Benchmarking: Service and Performance Indicators**
(*Continued*)

Quality of Service
- Continuity of service as hours per day and days per week
- Percentage of customers with discontinuous supply
- Quality of water supplied: number of tests for residual chlorine (as percentage of norm)
- Quality of water supplied: samples passing on residual chlorine
- Complaints about services as percentage of water and wastewater connections

Wastewater treatment
- At least primary treatment as percentage of total wastewater
- Primary treatment only as percentage of total wastewater
- Secondary treatment or better as percentage of total wastewater

---

BOX 2.6 **Benchmarking: Efficiency and Financial Indicators**

Cost and Staffing
- Total annual operational expenses as a proportion of total volume sold, measured in US$ per cubic meter sold
- Operational expenses for water supply only, measured in US$ per cubic meter sold
- Percent of operating costs spent on water service
- Percent of operating costs spent on wastewater service
- Unit cost for wastewater service as US$ per person served
- Staff per 1,000 water connections
- Staff per 1,000 water and wastewater connections
- Staff per 1,000 water population served
- Staff per 1,000 wastewater population served
- Labor costs as a proportion of operational costs
- Energy costs as a proportion of operational costs
- Contracted-out service costs as a proportion of operational costs

Tariffs, Billing, and Collection
- Average tariff for services
- Total revenues per population served as a percentage of GDP
- Residential fixed charge

(*Continued*)

BOX 2.6 **Benchmarking: Efficiency and Financial Indicators**
(*Continued*)

- Ratio of industrial to residential charges
- Connection charge
- Collection period in days (accounts receivable/total annual billing times 365)

Financial Performance, Assets, and Investments
- Operating cost coverage ratio (annual operational revenues as a percentage of annual operating costs)
- Debt service ratio
- Gross fixed assets, water and wastewater, measured in US$ per total population served
- Gross fixed assets, water, measured in US$ per water population served
- Gross fixed assets, wastewater, measured in US$ per wastewater population served
- Total annual investments as a percentage of total annual operating revenues
- Total annual investments per person served (water)

Affordability
- Total revenues per service population as a ratio of gross national income per capita
- Annual water bill for a household consuming six cubic meters of water per month through a household or shared yard tap (but excluding the use of stand posts)
- Connection charge as a percentage of per capita gross national income

---

2

produces a roadmap for achieving the required improvement by looking at best practices in the sector. (For an illustration of process benchmarks, see box 2.7) Thus, metric and process benchmarking complement each other in an overall performance improvement program.

About a decade ago, the World Bank started to develop a set of indicators for water and sanitation utilities. This effort led to the creation of the International Benchmarking Network for Water and Sanitation Utilities (IBNET, *La Red Internacional de Comparaciones para Empresas de Agua y Saneamiento*), which is funded by the United Kingdom and operated in collaboration with the World Bank and the Water and Sanitation Program (WSP, *El Programa de Agua y Saneamiento*). IBNET has become an important resource for the benchmarking of water utilities. Its Web site (http://www.ib-net.org/), which is in English, Spanish,

2

BOX 2.7 **Benchmarking: Process Indicators**

What best describes the utility's planning process?
- Setting budgets for next year
- A multiyear plan that identifies targets and resources for change and improvement
- Neither of the above (describe)

Does the management of your utility undertake the following?
- Skills and training strategy for all staff
- Annual appraisal and target setting system for managers
- Annual appraisal and target setting system for all staff
- Reward and recognition program for all staff
- Ability to recruit and dismiss staff (within an agreed plan)

Who has general oversight of the utility's services and prices?
- Local, regional, or national government department
- Independent board of stakeholders
- Independent service and price regulator
- Other (describe)

What are the main sources of finance for investment?
- Grants or government transfers to the utility?
- Borrowing from international financial institutions (multilaterals or bilaterals)?
- Government-owned banks?
- Commercial banks or bond holders?

Does the utility offer service and payment choices to its customers?
- More than one level of service for household or shared water supplies?
- More than one level of sanitation or sewerage service/technology for households?
- A flexible/amortized repayment option to spread the costs of connection to the water and/or sanitation network?

How does the utility find out the views of its customers?
- Letters and telephone calls from customers
- Inviting customers' views through radio, TV, or other publicity
- Questionnaire survey
- Other (describe)

and Russian, not only provides a detailed description of the methodology and definitions of the indicators but also gives data for hundreds of utilities around the world. This means that any utility that wants to start the benchmarking exercise quickly can make comparisons. Figures 2.1 and 2.2 illustrate the richness and usefulness of the data.

The value of benchmarks as an internal management tool is enhanced if key service-level indicators are disaggregated on a neighborhood and service-area basis. This will not only offer clear incentives to field managers to improve operations but also provide a better basis for investment decisions to ensure that the poor and other disadvantaged groups are served.

To increase transparency and accountability to the customers and taxpayers, the achievements should be made available to the public through various means, such as being displayed on the utility's Web site and reported in the press.

4.5   PROOF: The Public Record of Operations and Finance

Performance audits and quarterly financial statements are universally acknowledged as essential mechanisms and criteria of and for progress. They are natural complements to the benchmarks and performance indicators discussed above. If the utility is well run—that is, if costs are low, clean water is available 24 hours a day, 7 days a week, all households who desire a house connection have one,

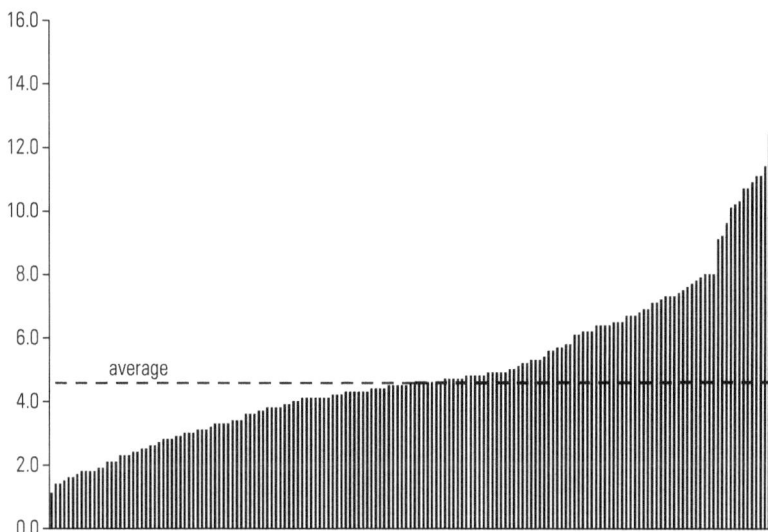

Figure 2.1   **Employees per 1,000 Connections (170 Latin American Water Utilities)**

Source: IBNET (http://www.ib-net.org/).

Figure 2.2 **Nonrevenue Water in Percent (120 Latin American Water Utilities)**

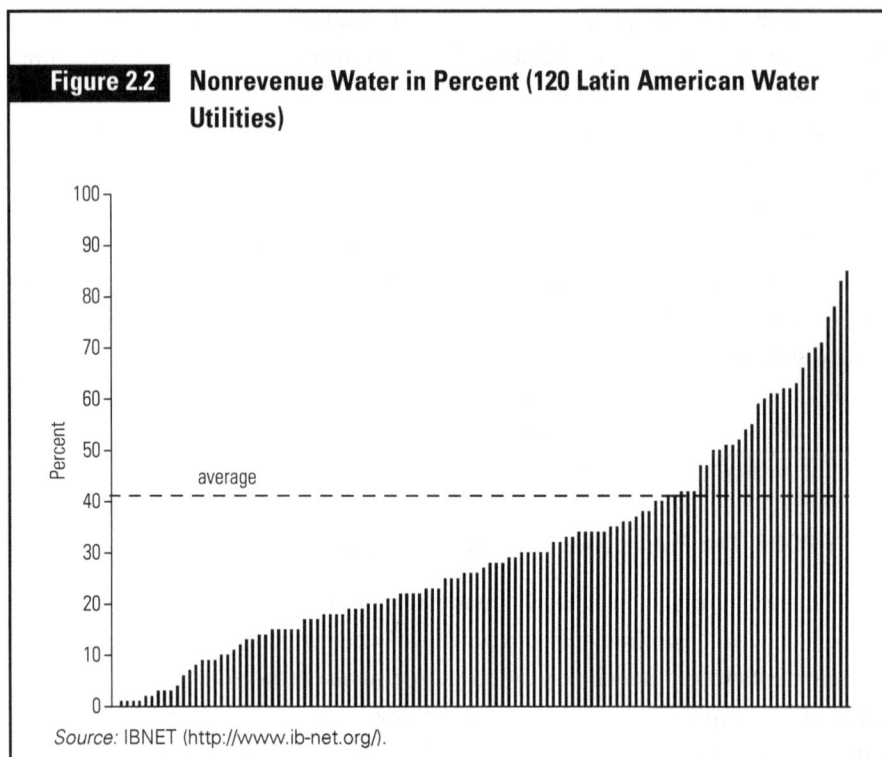

Source: IBNET (http://www.ib-net.org/).

everybody in the service area is connected to the sewerage system, and the wastewater is properly treated—the utility's management and the supervisory body responsible for the utility probably handle performance audits, budgets, and financial controls quite well. However, in most cases, increased public scrutiny and involvement in the budget process will help increase accountability and improve the allocation of resources.

With this objective in mind, four NGOs in cooperation with the local government in Bangalore launched what was called the "Public Record of Operations and Finance" (PROOF) campaign in 2002. The PROOF process involved making key budget documents and performance indicators public. These included:

- Quarterly revenue and expenditure statements compared with original budget figures
- Indicative balance sheet, with detailed information about current and long-term assets in addition to short and long term liabilities
- Key performance indicators for the service under consideration

These documents formed the basis for an informed and open discussion among public officials, NGOs, community groups, and interested citizens.

The PROOF campaign in Bangalore emphasized the sharing of full and accurate performance information. Each review served as an opportunity to bring financial accountability and performance into the public space. However, these reviews were also catalysts in a larger process of bringing the government and public closer together. Furthermore, each review also provided the basis for developing and reshaping public expenditure priorities. Bangalore was the city that pioneered the citizen report card that led to dramatic improvement in the water and sanitation sector, as described in box 2.1. Comparable improvements were also achieved in other public services. While neither the citizen report card nor the PROOF process can be given full credit for the upgrading of public services, the two tools were mutually reinforcing and helped to create a more open and responsive local government in the city.

2

Activity 2.4

Looking at the water utility in your town (or another company that you are familiar with), please answer the following questions:

Question 1: What are the strengths and weaknesses of the utility checklist?

Question 2: What are the strengths and weaknesses of the vulnerability assessment?

Question 3: What are the strengths and weaknesses of the performance indicators?

Question 4: What are the strengths and weaknesses of the PROOF: Public Record of Operation and Finance?

Question 5: Which methodology would you use for a detailed diagnosis of corruption and governance problems in your organization? Please explain your choice.

5   The Role of Regulatory Authorities

Regulatory authorities and other organizations with responsibility for sector over-sight play important roles in collecting and publishing data on utility performance. Irrespective of the regulatory formula adopted for tariff setting, the regulatory authorities are mandated to protect the interests of consumers and promote good governance in the water supply and sanitation sector. In short, the regulatory authorities face a trade-off between various objectives, as described in figure 2.3. Thus, regulators collect not only financial data but also performance and service quality data.

Most regulatory authorities tend to rely solely on technical (and financial) data provided by the water utilities themselves. These data may be similar to the performance benchmarks described in the previous section but are rarely as comprehensive. Furthermore, in most cases the data are not always accurate; in some cases the regulatory authority may be being bribed by the utility. In one major water utility in Asia, for example, nearly half of the flow and pressure gauges installed in the network were not in working order. The low-level staff responsible for reading and recording these data entered "estimates" rather than actual readings in the record books. Some data might be willfully misrepresented by linemen and meter readers. A service problem might be reported as "fixed" even if no action has been taken. Thus, there is a need for cross-checks. This can be achieved, to some extent, through internal control, monitoring, and audit systems. However, a lack of trust between the utility and the public is often a problem. People simply do not believe the data provided by the water company. Thus, it is also essential to get a view of service quality from the customer's perspective.

Figure 2.3   Regulatory Trade-Offs

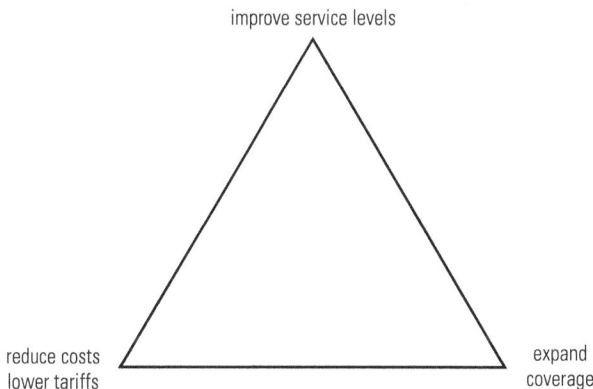

Source: Shordt, Stravato, and Dietvorst 2006.

As a regulatory tool, the citizen scorecard—which contains the assessment of the actual and potential customers—is, in principle, a suitable complement to the performance benchmark assessment. In practice, however, most citizen scorecards tend to focus on people's perceptions of the service rather than on more objective measures of service quality. Thus, if customer surveys are to serve as an independent verification of the utility's own reporting to the regulatory authority, there needs to be a good match between the utility's benchmarks and the questions asked in customer surveys. This is not very difficult if the survey questionnaire is designed with this objective in mind. Table 2.2 illustrates how the two types of data can be made more consistent.

The operating procedures for most regulatory authorities call for them to hold public hearings where consumers and other interested parties participate. To make these hearings productive and to increase the accountability of utility managers, all material related to utility performance should not only be made public but also be actively disseminated by the regulatory authority.

2

Table 2.2: Matching Utility Benchmarks and Customer Surveys

Performance area	Reported by utility	Customer survey
Network performance	Continuity of supply Water pressure Frequency of main breaks	Planned supply—hours/day Planned supply—days/week Frequency of interruptions
Water quality	Residual chlorine Coliform bacteria	Water smell Water taste Sand and other residual matter Water color
Service response	Frequency of billing problems Frequency of repair needs Time to repairs	Courtesy of staff Frequency of disputes Effectiveness of dispute resolution Frequency of problems Speed of repairs
Coverage	Percent with own water connection Percent less than 200 meters from standpipe Percent more than 200 meters from standpipe Percent with sewerage connection	Percent with own water connection Percent with well (own/neighbor) Percent from neighbor's tap Percent from tanker truck Percent from water vendor Percent less than 200 meters from standpipe Percent more than from standpipe Percent with sewerage connection

Source: Shordt, Stravato, and Dietvorst 2006.

Activity 2.5

Looking at your country, please answer the following questions:

Question 1: What role does the regulatory authority play in increasing transparency and promoting good governance in the water sector?

Question 2: Do you think the regulatory authority should directly address corruption in the water sector? Please explain why.

Question 3: What additional action could the regulatory authority take to address corruption and promote good governance in the water sector?

Question 4: In your opinion, who should have the main responsibility for addressing corruption and promoting good governance in the water sector: the central ministry in charge of water supply and sanitation, the regulatory authority, or the municipal government? Please explain why.

6   Concluding Activity

We hope that you have found the material in this module useful and that it has helped deepen your understanding of how to diagnose corruption. To give you an opportunity to once again think through the material, we suggest you complete the final activity below.

2

Activity 2.6

Assume that you have been empowered by the president to weed out corruption and improve governance in the water supply and sanitation sector. He wants to make this one of the cornerstones in his reelection campaign. You have only three years to produce tangible results. Fortunately, the budget is virtually unlimited.

Question 1: What information do you need to gather to set priorities and formulate an action plan?

Question 2: In this endeavor, with whom would you team up? Please explain your choice of partners.

Question 3: How successful do you think you would be? Please explain your optimism or pessimism.

Notes

1. This module is based primarily on TI and UN-HABITAT (2004) and WBI (2003; 2004).
2. Unaccounted-for water is simply the difference between the amount of water released into the system minus the amount billed to customers and used for other legitimate purposes (for example, water supplied to unmetered, free public standpipes). It is commonly expressed as a percentage of the total amount supplied.
3. If, for example, a large number of the water supply points are not metered, the reported amount of unaccounted-for water can be manipulated by changing the assumptions regarding legitimate but unmetered consumption.
4. See, for example, TI and UN-HABITAT (2004) and TI (2001).
5. For further information, see Davis (2004).

2

Tools for Addressing Corruption in the Water and Sanitation Sector

John Butterworth
Donal O'Leary
María González de Asís

Contents

MODULE 3

1 Introduction

1.1 Welcome

This is the third module of a course that is aimed at people engaged in or interested in the water and sanitation sector in Honduras and Nicaragua and focused on improving governance to address problems of corruption. Module 1 discussed the many potential types of corruption in the sector and explained that corruption is one of the reasons why people still fail to get access to the water and sanitation services they deserve, despite all the investments and improvements that continue to be made. Module 2 showed how it is possible to investigate the extent of corruption and the preparedness of service providers and other organizations to prevent it using both internal and externally focused diagnostic tools.

But what can you do, as an individual professional, about corruption in the water and sanitation sector? Module 3 focuses on the methodologies that can be used to improve transparency in decision making, enhance the accountability of officials and agencies, and improve the information available to citizens—all strategies that ultimately can help prevent corruption and its negative impacts on service delivery.

1.2 Goals of the Module

This module aims to present, accessibly, a range of existing tools for improving transparency, accountability, and access to information that you can use to help prevent corruption in water supply. The tools presented are not exhaustive, but we have selected those that are probably most relevant to the issues involved in municipal water supply and sanitation in Honduras and Nicaragua.

1.3 Learning Objectives

At the conclusion of the module, you will be:

- Able to match useful tools and approaches appropriate to specific situations you may encounter
- Aware of examples where simple tools were applied, with both positive and negative outcomes
- Aware of sources of further information on tools and approaches to help you develop your own action plans and activities

1.4 Outline of the Module

After a short discussion of different types of strategies and tools for improving access to information, transparency, and accountability within the context of municipal water and sanitation service delivery, some selected tools are summarized. The tools are each illustrated with short examples, and we also include advice on which tools to use in different situations. There are links to further source books where you can read about other tools or find further information about the tools included in this module.

1.5 Before We Start

Activity 3.1

We all have tools in our day-to-day work in the water and sanitation sector that can be used to help prevent corruption. Many of these tools are routine procedures for ensuring good performance of organizations and are standard good management practice.

Question 1: What are the existing tools and approaches you already have at your disposal to address corruption in water supply and sanitation? Take a few minutes to write these down.

Question 2: Do you make adequate use of these existing tools and approaches? If not, why?

Question 3: Do you need new tools and approaches? Why? For example, do you now need to tackle different types of corruption or to tackle it in innovative ways that might be more effective?

3

2   Identifying the Right Tools

This module draws heavily upon the toolkit, *Tools to Support Transparency in Local Governance*, developed by TI and UN-Habitat (2004), which advocates a broad approach to improved governance. Such an approach is also likely to be the most effective in combating corruption in water supply and sanitation. Although the tools for enhanced transparency, accountability, and access to information summarized here may be used alone and our focus is on the local level, in practice, you will need to use a combination of strategies and tools at different levels of governance. Identifying the roles and responsibilities of all key stakeholders in promoting good service delivery and making the most of their potential contribution is a good place to start.

This module also draws on the experience of other organizations such as the World Bank and the World Bank Institute (especially its Program of Open and Participatory Government at the Municipal Level, known as GAP for its Spanish acronym).

Key stakeholders in delivering water and sanitation services in Honduras and Nicaragua were identified in module 1. These include the national government and its agencies that have policy and regulatory roles (and that in some cases in both countries still operate water and sanitation services); local governments, which are taking on more responsibilities for operation of the systems, especially in Honduras; the private sector, which may be involved to various degrees from supplying materials and services to running concessions in a few cases; and nongovernmental and community-based organizations (NGOs and CBOs) that are especially important in rural water supply.

Three additional and important stakeholders are the media, which can play an important role in publicizing successes and failures; professional associations, which have a key role in promoting integrity; and citizens. Citizens of course have rights to basic services and water. Their voice as consumers is vital if utilities are to be held accountable, and in many situations individuals play key roles by to providing leadership in different contexts, lobbying bigger utilities, and taking active roles in community-managed organizations.

2.1   Types of Tools Available

Anticorruption tools can be classified in numerous ways. TI and UN-Habitat present four general strategic entry points as a framework for improving transparency in urban governance: assessment and monitoring; access to information; ethics and integrity; and institutional reform, as well as targeting of specific issues in the water and sanitation sector.

1. Promoting assessment and monitoring: This approach seeks to understand the types and scale of corruption and the degree of transparency in local

3

governance, while creating a baseline against which progress in improving transparency can be measured. This strategy is also valuable for increasing public awareness and mobilizing a constituency committed to tackling corruption. Monitoring in itself can start to reduce or prevent corruption. Related tools include surveys, appraisals, and report cards.

2. Promoting access to information: These are measures for improving stakeholders' access to information so that they may participate in decision making more effectively. Possible tools include holding meetings, passing laws guaranteeing access to information, using the media, and promoting public participation.

3. Promoting ethics and integrity: These are tools for clarifying what is expected from professionals and include monitoring mechanisms to ensure they adhere to their commitments and are sanctioned if they break public trust. These tools include conflict of interest laws, requirements for officials to disclose income and assets, codes of ethics, and integrity pacts.

4. Promoting institutional reforms: These reforms include the streamlining and simplification of administrative procedures and structural innovations to promote participation and accountability. These innovations may range from establishing complaints and ombudsman's offices to independent anticorruption agencies and the use of participatory budgeting.

The first of these strategies, assessment and monitoring of corruption, was covered in module 2. This diagnostic step is vital to identify the right strategies and tools. In this module, we focus on the second, third and fourth strategies.

Stålgren (2006) proposed a multipronged PACTIV approach for combating corruption in the water sector involving Political leadership, Accountability, Capacity, Transparency, Implementation, and Voice. Political leadership actions aim to mobilize support from political leaders and engage them as constructive anticorruption partners in water projects by demonstrating the potential political leverage from decreased corruption in the sector, including them at all stages in projects, and recording their commitments. Accountability actions aim to reform political and judicial institutions to reduce discretion and increase integrity; related actions might include checking contractors' support of political election campaigns and strengthening independent auditing. Capacity actions to strengthen public institutions and civil society could include increasing the technical competence of regulators and procurement officials, creating professional working environments with reasonable wages, and supporting independent data collection and diagnostics by civil society. Transparency actions aim to encourage openness and freedom of information and allow for advocacy and disclosure of illicit behavior. Possible tools include training media in investigative journalism on corruption in water, publicly displaying in newspapers and in villages information on water contracts and accounts, and disclosing water authorities' decision-making procedures and protocols. Implementation actions aim to put into action existing reforms and anticorruption tools

3

such as monitoring and the imposition of stiff judicial and economic sanctions on culprits. Voice actions aim to strengthen channels for water users, public officials, and private employees to voice discontent and report corruption and may include whistle-blower protection programs in utilities and public agencies and expanded voting rights in elections for water-related bodies.

2.2 Careful Design of Anticorruption Strategies

Given such a wide range of methods, and the complexity of corruption and governance issues in water and sanitation, the problem and its potential solutions seem daunting. Section 3 highlights some selected tools and gives examples on how they have been used successfully.

However, before proceeding it is worth reflecting on previous experiences that demonstrate the unintended and even harmful effects of worthy anticorruption activities (Stålgren 2006). More transparency can legitimize, and even increase, existing levels of corruption if it simply draws attention to corruption that is not condemned by the proper authorities and if the resulting punishment is perceived as negligible. Decentralization without proper support may multiply the number of potentially corrupt officials, and increased salaries may simply raise the corruption value of a public office. Corruption stamped out in one place may reappear elsewhere, and anticorruption actions to disband illegal service providers in slums may severely reduce access to services unless alternatives are provided. One key lesson is that diagnosis (the subject of module 2) is essential before planning any anticorruption actions or program. A second important lesson is that the effectiveness of any tools and actions must be monitored. Each country and situation is specific, and strategies and tools will, therefore, always have to be carefully adapted.

3

3 Selected Tools

This section introduces some key tools that can be used to improve access to information, transparency, and accountability. Each tool is illustrated by examples, and the objectives and key steps are set out. Links are provided to further sources of information and other tools.

3.1 Meetings to Develop Anticorruption Initiatives

Corruption is a difficult issue to acknowledge or address in the water and sanitation sector. You will probably need to start with very simple actions. Discussions and meetings to share experiences are usually the initial, and essential, steps for developing strategies to promote transparency, accountability, and access to information in the sector. However, these initial meetings and discussions can be difficult because the subject can inspire fear. Experience shows that people can be afraid, can disagree, and can even withdraw. People may express great interest and then not attend a discussion. Participants in a first meeting may try to compare the level of corruption in one region or country with another, offending other participants by doing so. People attending meetings sometimes do not, at first, want to give the name of the institution for which they work. In other cases, one or another particularly zealous participant may want to expose a person or company by name, which can disrupt any effort to share information and to have a positive experience.

This tool can be used to help hold successful meetings, especially a first meeting, in order to build trust and lay the basis for effective action by stakeholders to improve transparency and honesty and ultimately to reduce corruption. Because the subject is sensitive, extra care is needed to build trust through meetings and other events and communications. These suggestions and lessons learned, based on a paper by Shordt, Stravato, and Dietvorst (2006), may help you to work with colleagues and other professionals in the sector.

Be positive about the benefits of dealing with the subject and about what is known. If attendees are unfamiliar with each other, they should be introduced, but unlike meetings on other subjects, our experience is that participants may not want or need to mention their institutional affiliation until the end of the meeting, particularly if they do not yet know the other participants. Some ground rules need to be developed and explained at an introductory meeting: These might include, for example:

- Attendance should be voluntary. People should not normally be required to participate in a meeting on this subject.
- Begin by giving an overview of the plan of the meeting or workshop. This will inform participants about what to expect. Emphasize that the meeting will be practical, with the objective being, for example, to identify tools and strategies to enhance transparency and honesty.

- Remind the groups that transparency, honesty, and corruption are issues of importance around the world. No country is immune. However, this does not mean that corruption should be tolerated.
- Emphasize that there will be no particular "name calling," no exposures.

Think about the best ways to begin a meeting. Most people who attend a meeting on this topic have experience and ideas, which they can share. You may begin by asking each person to describe briefly why they have come. This can raise some interesting issues. One common observation is that, while the issue is important, perhaps the situation cannot improve until civil servants, in general, receive adequate salaries, or until the politicians and government in all sectors have improved. In other words, attendees may be asking: Can this subject be dealt with? Can we make a difference? Indeed, it is very difficult to ensure transparency and avoid corruption if professionals with very low pay are required to manage large resources and donor packages. The urge for rapid disbursement of funds is also an enemy of transparency and honesty in the sector. However, unless a beginning is made, there will never be an improvement, no matter how modest.

Does your group share a common language? An attempt to establish agreed-upon definitions of words such as honesty, corruption, and transparency can result in lengthy and not always productive discussion. An alternative is to "define" transparency, honesty, and corruption inductively. This involves listing and exchanging information about examples of opacity (lack of transparency) and corruption. For example, you might ask each person to make a list of all the different types of corruption or opacity that he or she can think of. The person who has the longest list can then be asked to read it out and other participants add more examples to the list. This can be enjoyable, fairly rapid, and informative. It can also lead to a common, agreed-upon understanding of the scope of the issues.

It is very useful to share information about tools or strategies that can reduce corruption and enhance transparency. Mainstreaming proven tools and strategies is a key to future improvement. Participants usually have some interesting examples, which deserve to be shared. This module contains a number of possible tools. It is often the case that several (although not all) strategies that can improve honesty or reduce corruption are not specifically directed toward this objective. Many of them are aimed at good governance and effective management, in general. However, the processes of good governance and management, in and of themselves, increase transparency and reduce corruption. Key questions to ask are:

- How can these tools and strategies be mainstreamed?
- How can they be scaled up?
- How can they be infused into projects and programs?

These types of questions, even though they are difficult, without immediately obvious solutions, still can lead to action. The discussions can be fruitful.

3

3.2 Access to Information Laws

Freedom of information is a basic right that is now enshrined in the constitutions or legislation of most countries.[1] These laws enable people to protect their rights (for example, to services like water and sanitation) and can be used by the public to guard against abuses, mismanagement, and corruption in government. But freedom of information can also help governments: more openness and transparency in the decision-making process can improve the trust of citizens.

According to a survey by the watchdog Privacy International in 2006 (Banisar 2006), nearly 70 countries around the world have put in place comprehensive freedom of information laws to facilitate access to government records. Another 50 had pending efforts at that time. The bulk of countries without active laws or plans to enact legislation are found in Africa, the Middle East, and Central Asia. There are some good examples in the developing world; for example, India's right to information law has teeth because civil servants can suffer financial penalties if they do not provide requested information (box 3.1). Many of these laws, however, are new and relatively untested, so their value in improving access to water and sanitation services is only now becoming apparent. Half of the countries with such laws have adopted their freedom of information legislation since 1996.

According to the IFEX (International Freedom of Expression Exchange), access to information laws have been enacted in several Latin America countries including Honduras, Belize, Colombia, Ecuador, Mexico, Panama, and Peru, while several other countries have bills in progress including Nicaragua.

In Honduras, the Transparency and Access to Public Information Law (*Ley de Transparencia y Acceso a la Información Pública*), was approved in November 2006 and published with amendments in December 2006.[2] Under the law, a National Institute for Access to Public Information (*Instituto Nacional de Acceso a la Información Pública*) is to be established as an independent body to process public requests for government information and oversee the law's implementation. Implementation was delayed for one year to allow government institutions and bodies to adapt to the law's requirements.

Freedom of information laws aim to make governments more accountable in their actions, such as service delivery. They do this by providing a right to access official records that enables people to scrutinize the performance of government agencies and officials. Key steps in ensuring effective access to information laws include:

- Adoption of freedom of information legislation. Most countries now have some relevant legislation.
- Making people aware of their rights to access government records. NGOs can play a useful role in helping vulnerable and marginalized groups, such as poor people, to benefit from such legislation.

BOX 3.1 **Using India's Right to Information Act to Enforce Rights to Water and Sanitation**

India adopted a right to information law in 2001. The citizens group Parivartan in Delhi has been active in supporting residents who use the legislation, including to seek improvements in water and sanitation services. Two examples cited by the group illustrate its effectiveness.

Three days after submitting a right to information application, a leaking water pipe was replaced in part of the city (Pandav Nagar). The pipe had leaked since it was laid, but previous complaints had been to no avail. The residents asked in their application about the status of their earlier complaints, the names of officials dealing with those complaints, the contract for the laying of the pipeline, the completion certificate for the works, and names of officials who issued it.

In another part of the city, a slum colony called Sundernagari in East Delhi, residents had struggled for 20 years to get sewers laid without success. As a result, people relied upon public toilets and women faced particular problems. In 2002, a social activist made an application under the Right to Information Act seeking information on the sewer system and asking when it would be functioning in their area. Progress was promised with surveys and tendering for work, but after a year nothing had been done. An additional application under the act allowed another citizen to inspect the files, but he was told there were no files. After protesting, officials admitted there was no work going on. Reminded that the earlier provision of false information could lead to deductions in their salary, the officials reacted by undertaking the survey work, and after residents lobbied the chief minister (the head of Delhi state) armed with information at their disposal, expenditure for the sewage system was approved and contracts were awarded for the work.

Source: Parivartan, www.parivartan.com (English only).

3

- Encouraging citizens to use freedom of information legislation to access government records. Fees may be charged and can be a barrier to use. Other restrictions that hinder access may include the need to produce ID cards.
- Seeking redress if a request for information is not dealt with properly under the law; officials and agencies may be liable to punishment for not making information available or for providing false information.

Increased access to information enables citizens to scrutinize the work of government, and more transparency can put pressure on government officials to be accountable, perform better, and shun corruption. The media has a key role to play in making information available that throws light on the performance of government and its agencies.

Activity 3.2

Access to information laws can be used to obtain information in the water and sanitation sector.

Question 1: What kinds of water and sanitation information do you think could potentially be made available using freedom of information laws in your own country (for example, performance indicators, tariffs, details of procurement processes)?

Question 2: Which organizations would need to provide this information (utility, regulator, local government)?

3

Question 3: By what mechanisms could information be better provided (publication in newspapers, radio, Internet, public hearings, other)?

3.3   Community Participation

Many of the tools that can be used to tackle corruption in the water and sanitation sector constitute established good practice (Shordt, Stravato, and Dietvorst 2006). Participatory methods and approaches are a good example. Methodologies such as participatory appraisal and community mapping are widely used in the sector. Community management, with high levels of citizen participation, has become the main model for service delivery in rural areas and has been adapted in some peri-urban contexts. There are many reasons for involving local citizens and community groups in planning, implementation and management, and decision making relating to service delivery. One of them is that the consistent application of known and tested participatory approaches can help to ensure transparency, promote honesty and best practice, and reduce corruption. Good use of these tools can help to avoid opening the door to corruption.

Participatory methods and approaches do not have an entirely good name everywhere, however. There is a tendency to subvert participatory approaches and tools to use them like mantras or formulas to collect information for its own sake, rather than use of the information locally to empower communities and groups to improve their access to water. Participation does not mean just collecting a lot of data from people and then doing a separate "expert" analysis. Equally, community management has earned a bad name in places where communities are left on their own to run water supply systems. Evidence shows that communities require support from local government or other support agencies if they are going to run sustainable water supply systems (Schouten and Moriarty 2003).

Participatory approaches can be used to harness the power of communities, or consumers, in planning or overseeing the delivery of water and sanitation services. By definition, if used properly, these approaches improve access to information and encourage transparency, and they can be very effective in limiting opportunities for corruption. Table 3.1 identifies some of the common problems encountered in water supply systems and areas where participatory methods can be used to good effect. These tools and strategies, if applied as intended, limit the scope for corruption.

3.4   Mapping and Site Selection

One important strategy to improve transparency and reduce corruption involves site selection and geographically based monitoring.[3] Proper site selection and monitoring are central to ensuring coverage while controlling cost. Site selection refers to identifying the physical location of water points, which should be done transparently, for example, with user and community involvement. Geographically based monitoring usually involves some type of mapping of existing water points and the location of people with service and without.

Site selection with the participation of local government and householders is an element of good management. With accurate maps and locally selected and

3

3

Table 3.1: Potential Uses of Participatory Methods

Common problems with water supplies	Why does this happen?	Which participatory tools can be used to help avoid this?
Facilities in the wrong place: Communal water points (such as tapstands in peri-urban areas) are not accessible for some users, especially the poor.	• More powerful people co-opt services • It is more expensive to build off the road where poorer people tend to live	• Mapping with participation of householders • Using GIS and GPS linked to mapping exercises • *Result: The poor cannot be overlooked*
Facilities never worked right: Construction water points, communal or household, just never worked properly.	• Bad construction • Bad planning	• Community participation in checking ongoing construction such as laying of pipelines for distribution networks or in rural areas, checking depth of well-drilling. Requires some training and organization of community groups. • Third-party monitoring of works before they are signed off and last payment is made. The community can also be consulted at this stage. • *Result: Better construction according to planned specifications.*
Irregular service: Water not supplied continuously or at regular periods.	• Inadequate or underdesigned systems • Poor operation and maintenance defective	• Put in place an iterative complaints system where users or local governments report faults to service provider; complaints go to the next level if no or inadequate action is taken. This might be an ombudsman but could also include the media. • Do not forget to plan for fault-reporting systems when designing and costing water supply systems. • *Result: Better run and maintained systems.*

High levels of unaccounted-for water.	• Leaks in pipes and pipe connections or interlinking with old systems • Stolen water through illegal or unauthorized connections • Water not billed and not paid for because a bribe was paid instead	• Encourage reporting of leaks and take remedial action (see above). • *Result: Less unaccounted-for water and better cost recovery.*
Water quality problems: Tastes bad, looks bad, is unsafe.	• Old pipes, poor construction, intermittent operation of systems • Treatment plant does not operate as intended (no testing, chlorine dosage irregular)	• As for irregular service, ensure quality problems are reported and addressed. • Share testing results with communities through media or participatory meetings. • *Result: Safer water quality.*
Water is expensive for the poor: To access safe water and protect their health, the poor have to spend a large proportion of their income (and time).	• Too hard to pay for poor • Well-connected people and institutions like government offices or businesses do not pay • Service is not good enough so people do not pay	• Use cross-subsidization between places, sectors, consumption levels, or groups so that tariffs are affordable for the poor. • *Result: More affordable water supply services.*

3

approved sites for water points, contractors and engineering staff can have clear and detailed directions to follow for their implementation work, helping to reduce opportunities for corruption in construction. This helps ensure that those who are meant to be served really do obtain access to improved water services (boxes 3.2, 3.3). A good process helps ensure that poor households will benefit as intended. Transparency in the site selection also helps involve communities in monitoring the construction when schemes are being implemented.

Although it is difficult to generalize, some key elements that would apply especially to site selection in rural water supply are to:

- Determine indicators and criteria for site selection through consultation with small samples of people from each group that will be involved.
- Identify the habitations of different social or economic groups, given that demands for water may differ. Often, poor households are less conveniently located, for example, off the paved roads. Without good participation and mapping, it is possible to overlook populations in less accessible areas.

3

BOX 3.2 **Example: The Importance of Site Selection and Targeting as a Tool in Malawi**

Some of the strategies that enhance transparency and reduce corruption are primarily focused on other issues, such as improved design and management. This is true for site selection and the monitoring of physical access to water points. Site selection refers to identifying the physical location of water points, which should be done transparently, for example, with user and community involvement. Monitoring usually involves some type of mapping of existing water points. Site selection and monitoring are central to ensuring coverage while controlling costs.

For example, the NGO WaterAid monitored the physical access to water points in Malawi using household surveys and global positioning system technologies. It found that the targeting of resources had not been pro-poor. The work showed that "if the resource allocation is effectively targeted at the unserved areas, the [UN Millennium Development Goals for water and sanitation] would be achievable even if the level of investment fell to 30 percent of what it has been over the last five years." Unfortunately a lack of targeting meant that some locations (usually more accessible and richer communities by the main roads) were served over and over again by rigs drilling new wells, while more remote and poorer communities were neglected.

Source: Shordt, van Wijk, Brikke, and Hesselbarth 2004.

BOX 3.3 **Example: Improving Transparency at Large Facilities, Kerala, India**

Four large water systems in the state of Kerala in India involved populations of 115,000 to 330,000 in rural, peri-urban, and small-town schemes. These were piped schemes with public water points, about half of which had already been constructed. As the construction was proceeding, there were some complaints. A study was then undertaken to assess mapping of the areas and site selection with the active involvement of local governments and households, particularly women who collect water. The original plans stated that about 80 percent of the current population would be covered by the original designs (defined as the proportion of the population living within a 250-meter walking distance from a water point), but subsequent mapping and site selection with the community showed that the real coverage levels that would be achieved were much lower. It appeared that only 30–70 percent of the population would be covered because the original maps were not accurate and some of the planned water points shown on the maps had been shifted to other locations. However, improved maps and the selection of better locations for public standposts showed that coverage could be increased by 20–40 percent while reducing per capita costs of the schemes at the same time. Some unnecessary water points were eliminated in negotiation with local government, community members, and local politicians. The cost of the whole site selection exercise was about 10 cents per person served.

The new site selection with community input meant a redesign of some parts of the distribution networks and some additional work, increasing the total costs by 22 percent. This amount would have been less if the schemes had been designed with better mapping and public site selection from the beginning. The communities, and particularly the poorer households, that benefited from the public mapping and site selection process were very enthusiastic about being involved, and they understood what they were paying for because the facilities were shown on maps that they had helped develop. Cost recovery from local governments proved easier because they had approved each water point in public meetings. Communities and the water authority also had accurate maps for subsequent operation and maintenance.

Source: Shordt, Stravato, and van Daalen 2006.

3

- Train and orient. In each location, different stakeholders should know their roles and agree publicly with the site selection indicator and criteria. Households should be informed in advance about the time and day the site selection will occur.
- Map and select proposed sites with potential users. Potential users from nearby houses can walk along the roads and paths with members of the water

committee and a field worker making a map together to suggest possible sites. Together, they can make sure that the rules of site selection are followed. Women, in particular, may be asked to select the preferred sites and sign in agreement of the possible site.

- Consider the marginalized: people drawing the map should ask about the location of the households that are left out. This can also help control political pressure and establish trust among the community at the beginning of a program.
- Ensure consultation and agreement. When a plan is approved, public meetings may be held. Representatives of local government, the water committee, and users may meet to approve sites for the water points. They can use the maps to verify the location and check the number of households that are left out.

If little attention is given to having accurate maps and identifying water points and connections with user and community involvement, these are likely to become "owned" by more powerful families that may limit the amount of water available to other households. Gender issues are important in the location of water points because women and young people usually collect water and therefore should have a voice in deciding their location. Women are often better at mapping (they know more about their area) and selecting locations for water points (they carry water from the public taps).

3

Activity 3.3

Harnessing the power of communities and citizens to improve governance can be a useful tool.

Question 1: At what stages in the project management cycle of the utility you are familiar with could communities and citizens be involved in oversight and approaches to reduce graft and improve efficiency?

Question 2: Which activities in the project management cycle are not readily improved through community or citizen involvement?

3

Question 3: How might community or citizen involvement worsen corruption if it is not supported or done carefully?

3.5 Raising Citizens' Voice

A key gap in many situations is the inability of consumers to hold water and sanitation service providers accountable. Often citizens have no mechanisms or only weak ones for making legitimate complaints and seeking redress for the poor or expensive services they might receive. This is a particular problem given the monopolistic nature of water and sanitation service supply. Consumers do not usually have the option of withholding their business and using another provider because normally those do not exist. A number of organizations including the NGO WaterAid see the voice of citizens as a key missing link and an underlying reason for generally low performance in the sector.[4] It is one approach for improving social accountability in water and sanitation.

Citizen action as promoted by WaterAid helps communities (groups of consumers) to prepare to engage with service providers and government and then supports that engagement for as long as required (boxes 3.4, 3.5). Project partners facilitate the process, rather than mediating on behalf of citizens. In each initiative, local people develop a fuller understanding of their entitlements to water and sanitation, their current water and sanitation service situation, and the range of responsibilities for policy and service delivery. The approach piloted by

3

BOX 3.4   **Examples: Results of Citizens' Action Promoted by WaterAid in Asia**

In India, WaterAid reported that local people have had major public successes in areas such as freedom of information, right to water and making report cards, and are now developing their own forums for testimony and negotiation. Communities are also generating their own databases on the facilities in their area and their functionality, as well as the expenditure by service providers and cross-checking the reporting done by the authorities of service provision in their area. By publicly displaying the names and contact details of service provider staff, avenues are opening for citizens to question their performance based on the evidence in the database.

In Bangladesh, the central government provides a subsidy for the hard-core poor through the local government institutions to enable them to build their own latrines. Communities were concerned that richer households are mainly capturing these subsidies. Through community-based organizations, the villagers collected lists of households that had received government subsidies for sanitation over the past two years. Communities determined the possible flaws in the preparation of these lists as they knew best who was eligible for the subsidies in their areas.

(Continued)

BOX 3.4 **Examples: Results of Citizens' Action Promoted by WaterAid in Asia (*Continued*)**

In one area, through such an exercise, it was found that almost half of the subsidies distributed had gone to better-off households and not to the poor as intended

In Jahanabad Union, Rajshahi district, Bangladesh, the facilitating NGO, Village Education Resource Centre, prepared an *Odhikar Potro* (Rights Paper) that contains, in simple language, the water and sanitation policy entitlement of the people under various laws and policies and government's commitments to the people. The community was excited to see, read, and internalize this information and was inspired to engage in dialogue with service providers to realize their entitlements. In India, in the inaccessible Santhal Parganas region of Jharkand, the rural water supply scheme called Swajaldhara, and the Total Sanitation Campaign are run by the federal and state government. People's entitlements under these schemes were simplified and explained to the communities. Local youth were built up as community cadres and trained on the nuts and bolts of these schemes and how to disseminate information regarding them. They were also trained on other empowering legal provisions made by state and central government for accessing information and decision making, such as the government mandated social audit methods and the Right to Information Act of 2005.

Source: Swain, Wicken, and Ryan 2006.

3

BOX 3.5 **Example: Citizens' Voice in South Africa**

A program in South Africa, the "Raising Citizens' Voice Project" in Cape Town, also used a similar approach. There, the first step focused on building partnerships with the various stakeholders in the province and selected municipality. This collective buy-in was seen as critical for the smooth functioning of the initiative as well as for its sustainability over time. The second step was raising citizens' capacity to hold local government accountable, largely achieved through community training. The third step was institutional reform of the water department to enable it to be more responsive to the public input regarding services. The fourth step was institutionalizing a public oversight mechanism through which the public could play a public monitoring role.

WaterAid includes community mobilization, generating a picture of service levels, raising awareness of entitlements, preparing for engagement with providers, and a dialogue.

- Community mobilization: People are encouraged to become involved through their existing community organizations. Where these are weak, then the need for support is greater and longer.
- Generating a picture of service levels: Local people are assisted in choosing and then implementing a method for collecting and analyzing information about their water and sanitation services. These can be relatively structured methods, such as report cards, where local people rank or score the range of their services at a communal level, or less structured methods, such as forums for public testimony and sharing of experiences or juries of citizens that meet periodically to compare experiences and then move to discuss and demand changes.
- Raising awareness of entitlements to water and sanitation: Community members are assisted in more fully understanding their water and sanitation entitlements by right, law, or regulation.
- Preparing for engagement with providers: With the data they have collected, citizens can compare the service they actually receive with their entitlements. Training in negotiation is provided if needed. People can discuss how to approach service providers and what their objectives will be in any dialogue.
- Dialogue: Communities can start negotiations with those responsible for providing services or developing policy. Partners give support for as long as necessary.

3.6 Participatory Budgeting

Participatory budgeting is an innovative financing mechanism that gives citizens a bigger say in a key issue: how money is spent by government for local development. The process of participatory budgeting started in Porto Alegre in Brazil in 1989 (box 3.6) and with some considerable success has been implemented there and elsewhere throughout Brazil, as well as spreading to other Latin American countries, including Argentina, Bolivia, Colombia, Ecuador, Mexico, Peru, and Uruguay, and in Europe, Africa, and Eastern Europe (Menegat 2002).

Participatory budgeting aims to spend government monies more wisely by extensively involving citizens in the setting of investment priorities for public works and in the oversight of implementation.[5] It ultimately aims to empower local people in participatory government or participatory democracy. Although there are differences between cities, the key steps in participatory budgeting as developed in Porto Alegre are:

- Development of forums in which citizens are able to control and steer the municipal government and its spending. Communities participate in assemblies

3

BOX 3.6 **Participatory Budgeting in Brazil**

Example: Belo Horizonte
Belo Horizonte, one of Brazil's larger cities, has over 2 million people and 160 favelas (informal settlements). The city's governance structure is divided into nine regional authorities, who are appointed by a single municipal authority, run by the mayor. In 1993, a new government that had run on a pro-poor platform was elected. To make good on its promises, the new government adopted a participatory budgeting approach to municipal finance as a means to increase transparency and accountability within city government and to engage and encourage participation by citizens and community groups.

Through the system of participatory budgeting, the regional authorities were further subdivided by population and physical boundaries to encourage participation at a very local level. Administrators at the regional authority were tasked with providing information about the city's finances and administrative functions and with guaranteeing citizens' rights to define government goals and strategies to achieve social needs. Participating citizens and community groups were tasked with defining local investment needs.

Although the first year of participatory budgeting faced some challenges, particularly in generating participation, the response in the second year was intensive action and engagement by the regional administrators; adaptation and acclimation by municipal authorities to the approach; and even greater responsibility delegated to citizen control. Participation in the second year increased by 80 percent. According to a 1994 Gallup opinion poll, city residents perceived the new government's key accomplishment to be the participatory budgeting process, with a wide majority supporting the government's outreach efforts and clarity in explaining the city's budget. In 1994, $15.6 million, or 40 percent of Belo Horizonte's total investment budget, was earmarked for participatory budgeting, allocated among the nine regions. What resulted was a shift in municipal funding primarily toward sanitation and basic infrastructure (including roads), followed by funding for site preparation for additional water, sanitation, drainage, roads, and other public assets. This was particularly so in the favelas, where investment also switched from large-scale capital works that had limited direct impact on the poor to ones that had a clear impact.

Over time, participation has increased as the government continues to demonstrate its ability to respond to citizen demand through investment and through a clear framework for monitoring results. The process is highly transparent, and government officials have developed skill at predicting and responding to issues as they arise. State-level officials have had to shift their approach and become more open, providing timely information to the regional authorities on request, in order to inform local decision making. At the same time, technical and financial constraints mean that not all citizen demands can be met, and detailing these constraints

(Continued)

3

BOX 3.6   Participatory Budgeting in Brazil (*Continued*)

requires patience and respect for citizens. By practicing transparency, citizens are more aware of their rights and, importantly, their obligations to the public sector, which has increased overall confidence in government.

Belo Horizonte's experience with participatory budgeting suggests a few factors that can lead to success, including political will and champions at all levels of the city's government to implement the approach; the existence of regional authorities within the municipal structure, which could extend outreach to very local levels; and a transparent process for allocating resources. Perhaps most important, the process is only as good as the follow-through—in this case a demonstrable ability to allocate public resources according to the priorities defined by the process. Some challenges still remain, particularly in communicating in a digestible way so that all citizens can understand city finances, including taxes and fiscal policy, revenue collection, and management.

Example: Porto Alegre and Participation in the Running of a Public Water Utility

The public utility (DMAE) that supplies water and sanitation services in Porto Alegre, Brazil, is financed through a progressive tariff that generates a surplus of 15–25 percent each year. Citizens use participatory mechanisms to propose and vote on new investments to spend this surplus. They are also represented on a citizen's board that oversees the utility and its contractors, promoting accountability. Citizens are therefore involved in both planning and oversight of DMAE's water services.

A range of methods are used to facilitate information sharing. Citizens can reach DMAE by phone, and the utility runs a Web site that includes information on planned projects and allows citizens to check their monthly consumption. The utility also has offices throughout the city where citizens can pay their bills or make complaints. These offices are particularly useful for low-income users who may not have access to other communication means.

Before participatory budgeting and citizens' board were put into place, DMAE primarily served business districts and affluent residential areas. Since citizen participation increased, DMAE has kept up with population growth and expanded services significantly. Between 1990 and 1995, the number of households served by the drinking water network in Porto Alegre was expanded from 400,000 to 465,000, with 98 percent of households being connected. In 1989, only 46 percent of the population had sewer connections, but this was nearly doubled to 85 percent by 1995. Over 10 years, public works totaling more than $700 million were implemented through participatory budgeting in Porto Alegre with the highest priority being basic sanitation. A key ingredient in the success of the Porto Alegre approach was the commitment of all stakeholders to the financial viability of DMAE.

Source: Bretas (1996), as summarized by Trémolet and others (2007); Smith (2006); and

organized by geographical district and sectoral theme to determine their needs and priorities. In addition to defining the municipal budget, communities also manage the implementation and timing of the public interventions.

• When both the communities' district and sectoral priorities and the government's own requirements have been established, a proposal is drawn up to be discussed with the Participatory Budgeting Council. Once approved, the budget proposal is sent to the city councilors. In the meantime, the Participatory Budgeting Council and the municipal government begin drawing up an expenditure plan based on the budget proposal. The expenditure plan sets out all the public works to be carried out in each district for that year and the government authorities responsible for their execution; the plan is printed and distributed to the public.

3.7   Access to Budget, Expenditure, and Performance Information

Regular reports and the accounts of a service provider provide an obvious, but valuable tool for improving access to information, transparency, and accountability. However, many municipal water and sanitation providers do not produce annually audited reports and accounts, even when they are required to do so. And frequently such reports are not widely accessible to anyone other than regulators within government. Although such reports are unlikely to be read by the majority of consumers, simple summaries can be provided with utility bills, and the information provided can be more widely distributed by the media and nongovernmental organizations. At the local level, information on projects and expenditures can be disseminated using simple but innovative means such as sign boards near community centers or painted notices on the infrastructure itself detailing the use of financial resources.

The objective of this tool is to provide regular information on budgets, expenditure, and performance of water and sanitation service providers. Key steps are:

• To produce regular audited reports and accounts in accordance within the requirements of sector regulations
• To provide key performance and other indicators, including the number of households with water and sewerage connections (and the number and characteristics of households without access), the cost of water and sanitation services, the number of staff employed by the provider per 1,000 connections, and the numbers of complaints and effectiveness in dealing with complaints
• To make reports and accounts widely accessible to the media and consumers.

3.8   Public Expenditure Tracking (PET)

Public expenditure tracking, or PET, aims to track the flow of public funds and other resources from the central government through the administrative hierarchy

3

and out to frontline service providers such as municipalities or utilities providing water and sanitation services. The key question that PET sets out to answer is: Did public funds end up where they were supposed to? (For an example, see box 3.7.)

To track expenditures in water and sanitation schemes, a possible series of steps would include:

- Identifying a small sample of projects or systems, ideally by random sampling, avoiding bias in making the selection. Describe the projects, technology used, and other related activities (such as hygiene promotion, community mobilization, joint planning, special participatory activities, and the like).
- Analyzing how much money (or materials, or both) was sent from the center, identifying the amount and purpose of resources provided from the central or regional government to local governments or service providers responsible for project implementation.
- Estimating or calculating the amount made available at the lowest level and used for the agreed purposes.
- Calculating a ratio showing the proportion of finance reaching the lowest level by dividing the estimate of spending at the local level by resources provided from the center.
- Analyzing any evidence that funds have been diverted (suggested by a low ratio of funds being spent locally) and why that might have happened. If significant funds have been diverted, identify the consequences, such as unserved or underserved poor households, inferior construction, irregular service, poor quality water, or disenchanted segments of the local population. Consider what tools or strategies could improve the situation.
- Analyzing the number of desks that paperwork crosses in implementing a water and sanitation project can also be instructive. If funds go through a large number of approvals, passing many desks, the chances of delay and misuse are greater.

BOX 3.7   **Example: Money Diverted from Education in Uganda**

One of the first PET surveys was in Uganda and tracked funds that were allocated from the central level for schools, mainly for construction. The study (Reinikka and Smith 2004) found that on average only 13 percent of the funds sent from the center arrived in schools, with 87 percent "captured" en route. A subsequent public transparency campaign to publish, post, and broadcast information about the financial grants led to a near reversal of those percentages, with 82 percent reaching the schools, a remarkable improvement. Schools near newspaper and communication points did better in receiving their funds.

3.9 Integrity Pacts and the Concept of Social Witness

The concept of the integrity pact was developed by Transparency International (TI and UN-HABITAT 2004) in the 1990s to help safeguard public procurement from corruption. The pacts can be used by a government agency or any other body in its procurement practice. Several countries, including Argentina, Colombia (box 3.8), and Mexico (box 3.9), have already implemented integrity pacts that cover infrastructure projects in the water and sanitation sector. The pacts are intended to reduce the high costs of corruption in public procurement, privatization, or licensing.

BOX 3.8 **Example: Selling Pipes with Integrity in Colombia**

In Colombia, self-regulation is being promoted through an integrity pact to reduce corruption in the procurement of pipes that are a key component of water and infrastructure projects. The procurement of pipes is vulnerable to practices like collusion in the tendering process and price-fixing that raises costs. Recognizing that such practices were a threat to their own business, private sector firms in Colombia, with support from Transparencia Internacional–Colombia and the government, took the initiative to regulate themselves by developing procedures to ensure transparent and fair procurement in the sector, developing indicators to monitor compliance, and establishing a sanctions and ethics committee to take action against transgressors.

Alma Rocio Balcazar of TI–Colombia reported that in their first year or so, these measures reduced prices by 30 percent. This money saved was previously going to sales agents who used it to pay bribes and to influence which firms were awarded tenders. Half of the 167 distributors in the industry had signed the agreement, and the ethics committee has received its first reports. Although it is still in the early stages, the initiative is promoting cultural change within the companies that are now focusing in a very positive way on how to do clean business.

The sectoral antibribery agreement was initiated in Colombia by ACODAL—the Colombian Association of Environmental and Sanitary Engineers, whose affiliated water pipe manufacturing companies accounted for 95 percent of the national market and 100 percent of the bids in public tenders for water supply and sewer systems. ACODAL approached TI–Colombia, and the two organizations worked together to develop an agreement among the piping companies based on TI's Business Principles to Counteract Bribery (BPCB). A similar agreement was signed in Argentina in December 2005. Agreements are also being considered in Brazil and Mexico.

"The impact and effect of this Agreement will be very strong, since we have never before had a code to guide us on these matters. Now we have

(Continued)

3

3

| BOX 3.8 | Example: Selling Pipes with Integrity in Colombia (*Continued*) |

parameters for action. Furthermore, the sanctions that have been estab-lished are very important. With this Agreement we, pipe manufacturers, will act differently amongst ourselves, since the same rules and regula-tions apply to all" (Testimony of a participant in the agreement).

The agreement included the development of a general anticorruption pol-icy in each company as well as specific guidelines regarding each of the forms of bribery specified in the BPCB. The detailed guidelines covered issues such as pricing and purchasing, distribution and sales schemes, implementation mechanisms, internal controls and audits, human resources management, communications, internal reporting and consulting, as well as protection of "whistle blowers." In addition, the agreement specified the roles of an ethics committee and a working group tasked to supervise the implementation of the agreement and armed with far-reaching legal and economic powers that could be used against companies that failed to com-ply.

Lessons learned include:

- Ethical commitment and motivation can move private sector entre-preneurs to self-regulate and adopt common standards to reduce corruption.
- Leadership from top management of companies must be firm and enduring.
- Coordination with national governance reforms helps to mobilize political commitment, to move beyond the needs of specific individ-ual business and to ensure that the agreement is followed up by par-allel work in the public sector to prevent corruption risks arising from the state.
- Involvement of a third-part actor, such as Transparency International, can help coordinate and facilitate an agreement.

An integrity pact is a binding agreement between a procurement agency (usually government) and bidders for specific contracts. It enables companies to refrain from bribing by assuring them that their competitors will also refrain from paying bribes. Government agencies also pledge to undertake to prevent corruption, including not seeking those bribes. The essential elements are:

- A pact (contract) made between the government office inviting public tenders for a supply, construction, consultancy, or other service contract, or for the sale of government assets, or for a government license or concession (the authority or the "principal") and those companies submitting a tender (the "bidders").

BOX 3.9 **Example: Using Social Witnesses in Integrity Pacts in Mexico**

In Mexico, the social witness is a representative of civil society who acts as an external observer in the procurement process. This innovative practice to promote transparency, diminish the risk of corruption, and improve overall efficiency of procurement has been used for several years, following Transparencia Mexicana's recommendation. The social witness makes recommendations during and after the procurement process and provides public testimony.

The social witness is a highly honorable, recognized, and trusted public figure who is independent from the parties involved in the process. He or she has full access to the relevant information and documentation and also has the right to participate in critical stages of the procurement process, especially:

- Checking the basis of the bid and the bidding notice
- Observing all the sessions that are held with possible bidders to clarify any doubts they may have
- Receiving the unilateral integrity declarations from the parties
- Witnessing the delivery of technical and economic proposals
- Observing the session where the successful bidder is announced.

Regulations specify criteria for participation of the social witnesses in procurement, and a list of registered social witnesses is published on the Web site of the Ministry of Public Administration (www.funcionpublica.gob.mx/unaopspf/unaop1.htm). To obtain registration, social witnesses must:

- Not be public officials
- Have no criminal record nor have been sanctioned or disqualified
- Declare formally that they will not participate in a procurement that could lead to a conflict-of-interest situation (such as a family or personal relationship with a bidder or a business interest in the project)
- Have knowledge of legal regulations related to procurement (or be willing to attend a training session).

In case of failures in ethical behavior or disclosure of information in the procurement procedure, the social witness is liable to sanctions.

Transparencia Mexicana acted as the social witness for the procurement of sewerage treatment services by the municipality of Saltillo in 2004, a contract worth almost $5 million. The organization followed each stage of the procurement process, attended meetings, and provided advice to the municipality. It produced a signed summary statement on completion of the procurement testifying that the process was proper and explaining what happened at different stages (for example, why certain bidders failed) and why the contract was awarded to the successful bid-

3

- An undertaking by the principal that its officials will not demand or accept any bribes, gifts, or other favors, with appropriate disciplinary or criminal sanctions in case of violation.
- A statement by each bidder that it has not paid, and will not pay, any bribes.
- An undertaking by each bidder to disclose all payments made in connection with the contract in question to anybody (including to agents and other middlemen as well as to family members or friends of officials); the disclosure would be made either at the time of tender submission or upon demand of the principal, especially when that bidder is suspected of a violation.
- The explicit acceptance by each bidder that the no-bribery commitment and the disclosure obligation, as well as the attendant sanctions, remain in force for the winning bidder until the contract has been fully executed.
- Undertakings on behalf of a bidding company will be made in the name and on behalf of the company's chief executive officer.
- A preannounced set of sanctions for any violation of its commitments or undertakings by a bidder, including denial or loss of contract, forfeiture of the bid security and performance bond, liability for damages to the principal and the competing bidders, and debarment of the violator by the principal for an appropriate period of time. Some or all of these sanctions may be taken.

Bidders are also advised to have a company code of conduct clearly rejecting the use of bribes and other unethical behavior, as well as a compliance program for implementing the code of conduct throughout the company.

3.10 International Conventions

Like other global issues, tackling corruption requires internationally agreed solutions and local action. To improve governance and reduce corruption, at least 12 international conventions and guidelines, and at least 7 donor policies have been prepared, largely over the past decade (Shordt, Stravato, and Dietvorst 2006). However, the implementation record for these conventions and policies is still somewhat disappointing. Many observers argue that the enforcement of these policies and conventions in each nation is a key global challenge (Swardt 2005). Some also call on the multinational and bilateral donors to work harder to implement their own policies in deed as well as word (Bailey 2003).

Important conventions in the Americas regarding corruption are the United Nations (UN) Convention against Corruption, the UN Transnational Organized Crime Convention, the Organisation for Co-operation and Development (OECD) Anti-Bribery Convention, and the Organization of American States (OAS) Inter-American Convention Against Corruption (box 3.10).[6] Whether they take a comprehensive or a more selective approach, international agreements or conventions provide a framework of rules and standards that facilitate international cooperation;

BOX 3.10 **Example: Anticorruption Conventions in the Americas**

The Inter-American Convention against Corruption (IACAC) of the Organization of American States was the first international judicial instrument dedicated to fighting corruption. It entered into force in 1996 and was ratified by a large number of American countries, including Honduras in 1998 and Nicaragua in 1999.

The United Nations Convention against Corruption is the first global convention against corruption. It entered into force in 2005 with the signature of 148 countries and 80 ratifications, including Nicaragua and Honduras. It obliges the parties to implement a wide and detailed range of anticorruption measures affecting their laws, institutions, and practices in order to promote the prevention, detection, and sanctioning of corruption, as well as to promote cooperation between countries. These include mechanisms to prevent corruption and repress certain corrupt practices, including prosecution when the practices involve activities such as bribery, passive bribery, transnational bribery, illicit enrichment, the improper use of classified or confidential information, the improper use of state property, the use of influence on public authorities for illicit personal gain, and the diversion of property or assets.

Although initially lacking, both the UN convention and the OAS convention now include monitoring and follow-up processes. Civil society organizations play a key role in augmenting monitoring. Examples include the establishment of follow-up commissions giving feedback to the convention review process (*Poder Ciudadano* in Argentina); a television program on the anticorruption conventions (*Chile Transparente* in Chile); training of civil society in Central America in follow-up mechanisms (*Transparencia por Colombia* in Colombia); programs to monitor the implementation of OAS conventions (*Transparencia Venezuela* in R. B. de Venezuela and *Acción Ciudadanía* in Guatemala); and citizens guides to the convention and training program to promote citizen participation in combating corruption (*Probidad* in El Salvador and *Cooperación Latinoamericana de Desarrollo* in

---

3

provide a checklist for reforming governments; establish a basis for governments to monitor one another; and represent a tool for civil society groups to use to hold their governments accountable. They can provide an important stimulus for the local action that is ultimately needed to improve transparency, accountability, and access to information.

The principal procedural steps associated with these type of conventions include negotiation, adoption, signature, ratification to accession, deposit of instrument of ratification or accession, entry into force, conference of state parties, implementation into law, implementation into institutional plans and procedures, and application and enforcement. None of the Central American countries is yet at the stage of fully implementing its conventions regarding corruption and passing it through all institutional procedures.

Conventions may include:

- Preventive measures to create conditions that promote good, honest, transparent, and efficient public management and high private sector standards of behavior; public sector ethics and procedures; public procurement; public sector finance; public reporting, access to information, whistle-blower protection; public education; private sector standards, including accounting and auditing standards; as well as measures to prevent money laundering
- Punitive measures calling for punishment of corrupt actions by means of judicial or administrative bodies, with the adoption of the necessary legislation and other measures to establish these as criminal offenses under domestic laws
- International cooperation between law enforcement authorities to make cross-border law enforcement efforts more effective; measures include extradition; mutual legal assistance in investigations, prosecutions, and judicial proceedings; and law enforcement cooperation, including joint investigations and special investigative techniques
- Processes for recovering illegally obtained assets, including investigative measures to trace assets, preventive measures to immobilize the assets (freezing, seizing), and confiscation.

3

While international political will is an essential piece of the puzzle, it is important that instruments at this level are created with sufficient "teeth" to make their implementation meaningful.

Activity 3.4

Ten possible tools for improving access to information, accountability, and transparency in water and sanitation service delivery are introduced in this module. From the perspective of your own organization, identify the strengths and weaknesses you see with each of these tools.

Tool	Strengths	Weaknesses
1. Meetings		
2. Access to information laws		
3. Community participation		
4. Raising citizens' voice		
5. Promoting consumer accountability		
6. Participatory budgeting		
7. Access to budget and expenditure information		
8. Public expenditure tracking		
9. Integrity pacts		
10. International conventions		

3.11 Public Meetings

3

Public meetings should be a normal function of legislators, civic officials, and other administrators of public services (like water and sanitation utilities) to provide information and to solicit the views of citizens.[7] Public meetings can serve to

- Ensure a better flow of information from public officials to citizens, especially about important decisions affecting them, and facilitate direct participation of stakeholders in local governance.
- Enable follow-up and public scrutiny of actions taken by public officials and government authorities, thereby increasing accountability.
- Foster better relationships between governments and local authorities and citizens.

The process of making information available to the general public, whether voluntarily or as a result of legal obligations, is evidence of a transparent government. Informed citizens are able to better advocate for accountability of public officials on their conduct and on decisions made on matters affecting the public such as service delivery.

In running public meetings, it is important to be clear about the issue and objective; to have a proposed agenda; to set dates well in advance, informing key actors and ensuring their participation; to notify the public of the meeting dates and agenda through print and electronic media, as well as through the Internet; to give options for the public to participate including by mail, fax, or e-mail in case they are unable to attend; and to provide contacts for further information.

3.12 Communications between Water Utilities and Their Consumers

In addition to the measures summarized in section 3.6, several other useful mechanisms can improve communication between water service providers and their customers.[8] These mechanisms include

- Providing performance-related information with bills
- Mounting information campaigns addressing investments made, coverage achieved, and quality of services
- Holding workshops to explain contents of contracts for investments, works, and service provision
- Publishing tariff structures
- Mounting campaigns on hygiene awareness and efficient use of water
- Conducting house-to-house surveys of users

Armed with more, and more useful, information, customers and organizations that represent them can act to ensure they get services they are entitled to at fair prices. Service providers in turn are better placed to provide services to meet demand.

3.13 Complaints and Ombudsman's Offices

It is widely known and accepted that more often than not, grievances and complaints about government bureaucracies, whether arising from the public or from within the organization, tend to fall on deaf ears. The legal systems in place in many countries that aim to address such grievances are often slow, expensive, and far from user-friendly. In addition, the courts of law may themselves be in disarray and subject to corruption, thus perverting the rule of law. Complaints and ombudsman's offices, which are sometimes separate but often combined, can provide an option for addressing such grievances within the local government system. The institution of ombudsman gives individuals an opportunity to place complaints about the practices of government or local authority before an independent and expert body, in addition or as an alternative to using the existing judicial system or internal complaints procedures. It is the independence of the ombudsman office that, above all, distinguishes such arrangements from other complaints procedures.

The primary function of the ombudsman is to examine a decision, process, recommendation, or act of omission or commission that appears to be:

- Contrary to law, rules, or regulations, or a departure from established practice or procedure
- Perverse, arbitrary or unreasonable, unjust, biased, oppressive, or discriminatory

3

- Based on irrelevant grounds or the exercise of powers or the failure or refusal to do so, for corrupt or improper motives such as bribery, jobbery, favoritism, nepotism, and administrative excesses.

Ombudsman offices also examine neglect, inattention, delay, incompetence, inefficiency, and ineptitude in the administration or discharge of duties and responsibilities.

3.14   Business Principles for Countering Bribery

Although most water and sanitation service providers are public agencies, the private sector provides a huge amount of related services and products. Transparency International's Business Principles for Countering Bribery (BPCB) provide a model for companies seeking to adopt a comprehensive antibribery program. Companies can use the principles as a starting point for developing their own antibribery programs or as a benchmark for existing ones.

To support the users of the principles, TI has produced a suite of tools, including a comprehensive guidance document, which provides additional background and practical information for those wishing to implement the BPCB or to review their own antibribery processes.[9] A six-step implementation process involves committing to a no-bribes policy; planning implementation of a program; developing program content including human resources policies, training, communications, and sanctions; implementing the program; monitoring, and evaluating and learning lessons. The development of an integrity pact within the Colombian water pipe industry was linked to implementation of BPCB (see section 3.8).

3.15   E-Government

The Internet has opened up new possibilities for governments and local authorities to interact with their citizens. Many local authorities, both in developed and developing countries, now run Web sites for their cities. Many cities also use the Internet to conduct as many transactions as possible with their citizens. Some countries are in the process of developing comprehensive "electronic government" or "e-government" policies and practices.

E-government makes use of the Internet to disseminate information. At its most basic, e-government requires a commitment by the local government or the organization that is placing information on an Internet Web page to maintain an up-to-date site. It thus requires human and financial resources as well as electronic capability on the part of the responsible organization. To make e-government work and for it to have a meaningful impact in a particular city, town or municipality, there should also be reasonably widespread computer literacy and access to the Internet for local residents.

3

BOX 3.11 **Example: Transparency in Public Procurement, Ecuador**

In Ecuador, the government does not readily offer access to information on public procurement. Responding to concerns expressed by businesses and other types of organizations about the high level of corruption in the country, a Web site was launched by the government in 2001 to provide information on public procurement. Anyone can visit the site and make inquiries about public procurement processes that are under way, completed, or pending approval.

As described by Sohail and Cavill (2007), the site provides information to its registered users who can search on criteria such as dates of tenders, name of organizations issuing tenders, location of projects, types of projects or services, and type of contract. For example, users can search to find all current bidding processes for water and sanitation projects. Each recorded entry provides more information, such as requirements for submitting bids, important dates, the amounts bid by tendering parties, and the origin of funds.

The project aims to demonstrate that making information available to the public leads to cost savings and transparency for state institutions. It also aims to highlight the various failings that make current systems of procurement bureaucratic and inefficient. Ultimately, it is hoped that this site will foster the development of better public services and will prompt the state to acknowledge that legislative reforms are necessary.

The e-government approach can be useful for providing general information, posting public notices on meetings, providing a place for the local community to report complaints, concerns, and emergencies, and to obtain various kinds of permits and licenses. One area of activity that is particularly prone to corruption and where Internet-based tools and databases have been used effectively is in public procurement (box 3.11).

3.16 Other Tools Identified by the GAP

A number of other additional tools, relevant to improving access to information, transparency, and accountability in water and sanitation service provision, are summarized in a paper by María González de Asís (2006). These include ethical campaign practices, citizens' charters, public waterdog groups, and anticorruption agencies.

4   Concluding Activity

In this module, 15 tools that might be used to help promote access to information, transparency, and accountability have been outlined. We learned that decisions about the actual tools used need to be based upon an assessment or diagnosis of corruption (see module 2) and that a wide range of possible strategies and tools are available, including tools not discussed here. We also learned that many of these tools can be used successfully only in specific circumstances or that they require specific skills or capacities to use them; often several different tools may need to be used by different stakeholders in water and sanitation for corruption to be tackled effectively.

Activity 3.5

Please take some time to think about the circumstances under which you might use these tools.

Question 1: Identify first steps that could be taken to use some of these tools and approaches. Focus on the ones you considered most relevant (in activity 4). Where could you start?

3

Question 2: What risks are involved in using these tools and initiating actions? Could your actions have unforeseen effects or make corruption worse? How could this be avoided?

Question 3: What could you do to safely promote anticorruption activities?

Notes

1. This section is based upon Transparency International (2006c); www.freedominfo.org/features/20060322.htm (in English only; a discussion of the role of freedom of information acts in development, including in Honduras and Nicaragua); www.privacyinternational.org/foi (in English only; the Freedom of Information Project of the watchdog Privacy International has produced a global survey of freedom of information regimes); and TI & UN-HABITAT 2004 (in English only).

2. According to International Freedom of Expression Exchange (www.ifex.org). Read the full law at http://www.se.gob.hn/content_htm/pdfs/leyes/ley.transparencia.pdf.

3. This section is based upon Shordt, Stravato, and van Daalen (2006).

4. This section is based on Swain, Wicken, and Ryan (2006).

5. For a general resource site (in English) on participatory budgeting and democracy, see www.participatorybudgeting.org. This site has links to many resources and organizations (some in Spanish).

6. See Transparency International, which provides links to a wide range of information on international conventions including follow-up measures; see www.transparency.org/global_priorities/international_conventions (including in Spanish). The UN Convention against Corruption is available in Spanish at www.unodc.org/unodc/crime_convention_corruption.html. Anti-Corruption Conventions in the Americas: What Civil Society Can Do to Make Them Work.

7. For more information on conducting meetings, see TI & UN-HABITAT (2004).

8. This section draws upon a presentation by Meike van Ginneken, of the World Bank, entitled "Mecanismos de rendición de cuentas a los usuarios de servicios de agua potable y saneamiento," June 2007, Global Development Learning Network (GDLN).

9. Produced by Transparency International and Social Accountability International. Available in English and Spanish at www.transparency.org/ global_priorities/private_sector/business_principles. The six-step implementation process is available in English and Spanish at www.transparency. org/global_priorities/private_ sector/business_principles/six_step_implementation_process.

Case Studies in Addressing Corruption in the Water and Sanitation Sector

John Butterworth
Donal O'Leary
María González de Asís

Contents

1 Introduction

1.1 Welcome

This is the fourth module of a course aimed at people engaged in or interested in the water and sanitation sector in Honduras and Nicaragua and focused on improving governance to address problems of corruption. We learned in module 1 about the many potential types of corruption in the sector and that corruption is one of the reasons why people still fail to get access to the water and sanitation services they deserve, despite all the investments and improvements that continue to be made. We learned in module 2 how it is possible to investigate the extent of corruption and the preparedness of service providers and other organizations to prevent it, using both internal and externally focused diagnostic tools. In module 3 we learned about a number of possible tools that can be used to improve transparency in decision making, enhance the accountability of officials and agencies, and improve the information available to citizens—all strategies that ultimately can help prevent corruption and its negative impacts on service delivery. This module introduces a series of case studies that can be used to explore in more detail some of the issues introduced in the first three modules.

1.2 Goals of the Module

This module presents international case studies drawn from Asia and Latin America that highlight experiences and lessons learned in other countries in applying successful strategies to improve access to information, transparency, and accountability in the water and sanitation center. Whenever possible, we have selected cases that are especially relevant to situations likely to be faced in Honduras and Nicaragua and that have not been widely documented elsewhere. The cases highlight good practices and the use of tools that are applicable to delivering water and sanitation in a wide range of contexts and that are relevant to organizations such as municipalities, water utilities, public-private partnerships, small-scale providers and suppliers, and regulators and support agencies. Each case study is supplemented with questions and exercises that can be used to unpack the issues in each case and introduce participants' own experiences and ideas.

1.3 Learning Objectives

At the conclusion of this module, the reader will be able to:

- Recall examples where appropriate tools were used to address different types of corruption

4

- Appreciate the necessity of using a range of tools in combination with the delivery of effective programs in promoting transparency and preventing corruption.
- Appreciate some of the process issues, risks, and unpredictable outcomes (both positive and negative) involved in anticorruption activities

1.4 Before We Start

Activity 4.1

Using Intervision to Tackle a Transparency, Information, or Corruption Problem

This exercise uses a method known as intervision (also called the incident method or peer assist) that is suitable for small groups to analyze problems such as a lack of transparency and corruption in water and sanitation. We suggest groups of five or six people. Following these steps will enable you to effectively deal with complex and unique problems, and enhance peer-to-peer learning by tapping everybody's resources. The exercise focuses on dialogue rather than discussion. It is about listening and learning—and talking only every once in a while. You will need about 45 minutes and you must try to stick to the time schedule. Talk only when it is your turn! Each group should appoint a timekeeper who also ensures that people wait for their turn before talking.

Before breaking into groups, make sure that a problem owner has been identified. The problem owner needs to have identified a clear problem that the organization currently faces. This could be any problem relating to transparency, access to information, or accountability issues within the organization, for example.

Then form small groups and follow these steps.

1. The owner explains the problem to the other members of the group without suggesting a solution (maximum 5 minutes).

2. Each of the other group members has a chance to ask clarifying questions to get a better understanding of the problem. The problem owner responds in turn to each question, but there is no discussion (maximum 10 minutes).

3. Each group member analyzes the case and the various aspects and gives her or his own solutions or ideas. The problem owner does not participate in the analysis, but just listens. There should be no discussion among the members of the group about each other's contribution (15 minutes).

4. The problem owner tells what he or she either likes or does not like about the

---

4

Table 4.1: Summary of Case Studies Included in This Module

Case study	Key characteristics	Key tools used to improve transparency, access to information, and accountability
Phnom Penh Water Supply Authority, Cambodia	Transformation of a public utility under difficult post-civil-war conditions; high levels of illegal connections and unaccounted-for water, including bribery of utility officials, effectively tackled	Best practice in human resources management; records management and computerization; complaint and reporting mechanisms; and reducing unaccounted-for water
Panama Canal Authority, Panama	Mainstreaming honesty and transparency as corporate values in the operation of a state-run navigation monopoly	Conflict-of-interest rules; disclosure of income and assets; code of ethics; and reporting of budget and expenditure information
Public Utilities Board, Singapore	A model public utility built around the integrity of staff to deliver efficient water and sanitation services	Best practice in human resources management; disclosure of income and assets; code of ethics; independent anticorruption agencies; complaint and reporting mechanisms; and one-stop shop for feedback

Source: Authors.

The case studies, their characteristics, and links to tools presented in other modules (and some new tools) are summarized in table 4.1.

2 Phnom Water Supply Authority: Cambodia

2.1 Introduction

Cambodia's Phnom Penh Water Supply Authority (PPWSA) is unlike a typical water utility in Asia. That is not because PPWSA has better service efficiency, greater water productivity, or an increasing consumer base, because other water utilities in the region have had some of these traits at one time or another. PPWSA is different because it has achieved all of these by radically transforming a decrepit and war-torn water supply system with missing water and missing customers into a model public sector water utility that provides drinking water to the capital city of Phnom Penh 24 hours a day (table 4.2).[1]

Cambodia's 20-year civil war and the era of Khmer Rouge rule destroyed much of Phnom Penh's buildings and infrastructure. The water supply system, whose capacity shrank from 155,000 cubic meters a day in the 1960s to 65,000 cubic meters by 1993, was deteriorating. With century-old pipes and a poor distribution network, roughly only a quarter of the city's 1 million residents received piped water.

PPWSA, the government-owned water supply utility, was having trouble meeting its challenges. Employees were demoralized, underpaid, and underqualified. Only 13 percent of connections had water meters, leading to inaccurate billing. Only 28 percent of water production was actually sold, with the collection rate not even reaching 50 percent. Much of the unaccounted-for water was tapped through

Table 4.2: Performance of the Phnom Penh Water Supply Authority, 1993 and 2006

Indicators	1993	2006
Staff per 1,000 connections	22	4
Production capacity	65,000 cubic meters/day	235,000 cubic meters/day
Nonrevenue water	72 percent	8 percent
Coverage area	25 percent	90 percent
Total connections	26,881	147,000
Metered coverage	13 percent	100 percent
Supply duration	10 hours/day	24 hours/day
Collection ratio	48 percent	99.9 percent
Total revenue	0.7 billion riels	34 billion riels
	(US$180,000)[a]	(US$8.7 million)
Financial situation	Heavy subsidy	Full cost recovery

Source: Asian Development Bank 2006.

a. At 2007 exchange rates.

illegal connections. Worse, PPWSA employees themselves were responsible for the water theft: they were installing illegal connections at a price of $1,000 per connection.

2.2 Turning Around Years of Deterioration and Neglect

The year 1993 marked the beginning of the restoration of Phnom Penh's water infrastructure. With the assistance of external funding agencies, particularly the Asian Development Bank (ADB), and through internal reforms, PPWSA transformed itself into an efficient, self-financed, autonomous organization in a city still recovering from long years of war and civil strife.

Ek Sonn Chan, the young engineer who took PPWSA's helm, initiated a "culture of change" within the organization, starting with the education and motivation of the utility's staff. This was followed by a flurry of reforms:

- The organization's workforce was streamlined and its morale improved. Higher management was given more responsibility, staff received higher salaries and more incentives, promising staff members were promoted, and a spirit of team-work was fostered.
- Collection levels were improved by installing meters for all connections, computerizing the billing system, updating the consumer base, and confronting high-ranking nonpayers and cutting off their water if they refused to pay, among other steps.
- The entire distribution network and treatment plants were rehabilitated. The utility hired locals instead of international consultants for the job and manually looked for the pipes because all blueprints had been destroyed during the civil war. It also mobilized the communities it served to report leaks and other problems.
- Illegal connections and unaccounted-for water were minimized by, for example, setting up inspection teams to stop illegal connections, penalizing those with illegal connections, and giving incentives to the public to report illegal connections.
- Water tariffs were increased to cover maintenance and operating costs. Tariffs were set to increase in three steps over seven years, but the third step was not required because the first two increases had produced enough revenues to covered costs.

The water service now covers 100 percent of inner city Phnom Penh and is being expanded to surrounding districts, with priority given to poor urban communities. In particular, PPWSA now serves 15,000 families in 123 urban poor communities by giving poor households extra privileges such as subsidized tariffs or connection fees, installment payments for connection fees, and more.

4

2.3 Lessons Learned

Here are some lessons from PPWSA's experience, identified by the Asian Development Bank.

Water Does Not Have to Be Free

The story of Phnom Penn demonstrates that a utility can charge for access to water and that the urban poor will be considerably better off paying for safe, piped water than they would be buying water of questionable quality from private vendors. For instance, Phnom Penh's unconnected residents used to pay 1,000 riels ($0.26) a day for water bought from private water vendors. Today, they only spend about 5,000 riels ($1.28) a month for PPWSA-supplied water.

Cost Recovery Is Vital

By developing a tariff structure where the costs of water production and transmission are fully recovered, the utility has become financially viable and is now able to invest in the water infrastructure. The PPWSA reached full cost recovery in 2004 and is now making modest profits.

The Operator Must Be Autonomous

Although the PPWSA is still government owned, it has enough autonomy to develop its own payment structure and culture, with an enthusiastic and motivated staff responsive to consumer demand, leading to more efficient operations and to the generation of revenues that could pay for infrastructure development.

4

Government Support Is Crucial

The tariff restructuring, which paved the way for PPWSA's greater revenues, would not have been possible without the support of the government of Cambodia and its development agencies. PPWSA would also not have had the freedom to innovate if the government had not declared the utility an autonomous body in 1986.

Civil Society Must Be Involved

The remarkable increase in bill collection and reduction in illegal connections also highlight the importance of involving users and civil society in a service that they want and are willing to pay for. The key has been to develop a utility-customer relationship, based on long-term community building rather than short-term contractual relationships. Effective awareness campaigns also enabled PPWSA to increase tariffs with broad public support.

Investing in Staff Yields Radical Results

Today, PPWSA takes pride in its team of people who are hardworking, responsible, and self-motivated. PPWSA professionalized its workforce, building its technical capacity (each staff member receives an average of 12 days of training each year) and instilling in its employees a work ethic of discipline, competence, and teamwork.

A Champion at the Helm Drives Reforms

When Ek Sonn Chan introduced the culture of change to PPWSA, he started the utility on the road to recovery. With each reform that PPWSA has taken, he has been its driving force, leading his staff and the community by example.

PPWSA has shown that clean water targets can be met through a transparent environment where water utilities have sufficient autonomy, where tariffs can cover costs, where service is equitable, and where there is the active involvement of staff and civil society. Ek Sonn Chan says, "It doesn't matter whether water distribution is done by the private sector or a public agency, as long as these institutions are transparent, independent from political pressures, and accountable."

In 2004, PPWSA was awarded the Asian Development Bank's Water Prize—an award conferred to exemplary project agencies that have established sound practice in implementing the bank's "Water for All" policy. Ek Sonn Chan also received the 2006 Ramon Magsaysay Award for Government Service for his "exemplary rehabilitation of a ruined public utility, bringing safe drinking water to a million people in Cambodia's capital city." The next challenge is improving Phnom Penh's sanitation system. Ek Sonn Chan says, "We convinced people to pay for the water they use. Now, we have to convince them to pay for the clean up of the waste they make."

4

Activity 4.2

Phnom Penh Water Supply Authority, Cambodia

1. We heard that previously water supply authority staff in Phnom Penh were receiving bribes for giving illegal connections to the system. Why might this have been happening? Consider the evidence from the case study and your own experience. Individually, suggest at least three possible reasons for the bribes, and write these down on cards (one reason per card). What strategies can you suggest to avoid such payments? Suggest at least three possible strategies and also write these down. Working in pairs, discuss your suggestions with a colleague.

2. Working in small groups of three or four people, imagine that you were working at the Phnom Penh Water Supply Authority in 1993 or a similar water supply system in crisis. Allocate some of the following roles to your group members: a) a new leader of the company who is supported by a development bank and the government to introduce far-reaching reforms, b) an experienced and senior manager at the company who has been making a nice living from taking a share of the bribes paid by consumers for illegal connections, c) an official from the development bank who wants to lend money to the utility to invest in its infrastructure, and d) a leader of a nongovernmental organization who is involved in campaigning for the rights of the urban poor. In your role, suggest what you think should be done. Make your arguments to other members of the group focusing on issues of access to information, transparency, and accountability.

4

3 Promoting Transparency in the Panama Canal Authority, the Largest Water Company in Panama

3.1 Introduction

Forty ships a day make the short, nine-hour journey along the Panama Canal from the Caribbean to the Pacific.[2] A ship sailing from New York to San Francisco through the canal travels 9,500 kilometers, well under half the distance of the previous 22,500 kilometer journey around Cape Horn. The Panama Canal was operated by the United States after its construction in 1914, but was eventually fully handed over to Panama in 1999. Each year more than 14,000 ships pass through the canal, carrying more than 203 million tons of cargo. In April 2006, a referendum in Panama authorized the canal's operator to undertake an investment program to increase the canal's capacity and to handle more and larger ships.

The Panama Canal Authority (*Autoridad del Canal de Panama,* or ACP) is a governmental organization established in the national constitution of the country and has operated since 1997 with the purpose of guaranteeing that the canal "may operate in a safe, continuous, efficient and profitable manner." Honesty and transparency count among ACP's corporate values and have made the authority—perhaps the biggest organization involved in the water sector in Central America—a leading organization in terms of its code of conduct and efforts to promote transparency.

3.2 ACP's Transparency Policy

The ACP has from the outset stated its vision to be a model of "excellence, integrity and transparency," and it does combine the theory and the practice in official documents (such as annual reports) as well as in a number of activities that counter opportunities for corruption.

4

3.3 Tools to Increase Transparency

The ACP uses three key tools to increase transparency:

- Publishing the salary and representation allowances of every employee of the authority
- Listing foreign trips by staff and their expenses
- Regulating the ethics and conduct of the ACP and its staff

The ACP has made all these mechanisms available for the public in its transparency page on its Web site.

The published list of salary and representation allowances offers a full overview for each of the 9,241 permanent and temporary workers at the ACP. It includes their position, salary scale, and basic wage as well as representation costs. This list is provided in a searchable database allowing access to all information fields; the date of the latest update is given to help provide an accurate account of personnel costs within the company.

This salary index is further supported by a list that tracks official foreign trips and their costs on an annual basis. The *reporte de viajes al exterior* shows the destination, purpose, duration, and cost of each trip. In addition to improving transparency, this list allows the ACP administration to make a more accurate estimate of the money spent in traveling.

However, what really constitutes the overall integrity and transparency framework for ACP's activities is the *reglamento de ética y conducta de la Autoridad del Canal de Panama* (regulations on ethics and conduct), also available on the Web site (figure 4.1). This 11-page document provides a general framework for a code of good conduct with clear definitions; overall ethical principles, particularly regulations for top executives; a guide for conducting internal activities and relations with external parties; and other information such as guidelines for the availability of reports.

The regulations focus on a number of typical transparency and integrity issues, including

- Conflicts of interest
- Abuse of position (in particular, at high levels) for economic (profit) or personal (unfair hiring of family or friends) ends
- Acceptance of gifts and preferential treatment

Figure 4.1   Employee Information Published by the ACP

Source: Panama Canal Authority Web site, Planilla de Empleados. https://apps.pancanal.com/ pls/defensoria/def2.p_inicio.

At all levels, the regulations set out a series of mechanisms for preventing misconduct, including

- Access to information and transparency in decision making
- Monitoring, early warning, and prevention measures
- Punishments for infringement of the regulations.

The ACP actively promotes transparency by providing key information and declarations such as salary and representation costs and costs of foreign trips as we saw above, through reports of annual accounts and financial figures, but also through the declaration of economic interests at the highest executive level to avoid future conflicts of interest. Early warning systems prevent conflicts of interest from happening. These include systems for voluntarily reporting potential conflicts of interest. Employees can accept gifts, but only with the condition that the gift is in line with the policy and interests of the authority and is acknowledged and accepted by the authority regulators.

On the other side of transparency, the ACP imposes restrictions on employees to prevent use of insider information in financial transactions. Similar restrictions also apply when an employee leaves the organization—employees are not allowed to use any information gathered under their supervision of a project at the ACP for two years after leaving the company.

Finally, the collegial and transparent decision-making process in the ACP helps prevent conflicts of interest. For example:

- Hiring of new staff is the responsibility of the ACP's managing director or his designate.
- Employees are required to get the approval from their supervisors before accepting external assignments.

The regulations focus particularly on gifts and preferential treatments that an employee or customer may enjoy. The regulations provide a maximum value for the gifts that employees can accept and impose restrictions on the conditions under which someone can receive a gift and for what reasons. Part of the monitoring in place at the ACP stresses that employees may decide by themselves how to deal with potential conflicts of interest: to agree to leave the issue to other employees, abandon his/her financial interest in the case at hand, or continue his duty but be stripped of regulatory authority. Among measures to prevent corruption and lack of transparency, the ACP imposes very strict working policies for relatives of an employee wishing to join the ACP.

If there is a breach of the transparency measures, in particular when it comes to appointing relatives to positions in the ACP, punishments, including loss of salary, come into play. Punishments are not the preferred way to ensure transparency at the

ACP, and usually employees are given an opportunity to change their behavior before being punished. The personnel manual on the one hand and the governing board on the other hand provide the checks and balances to make sure employees and top executives are respecting the code of conduct.

Activity 4.3

Management of the Panama Canal

A fish bowl is a special discussion format for group work. Form your group into two circles of people. The inner circle will discuss the following question while the people in the outer circle listen and remain silent. After a short while, the people in the two circles change places and the discussion continues (responding to the same question).

1. What are the characteristics of the ACP transparency framework that have most attracted your attention and why?

Group the ideas to make clusters of the most important characteristics and to simplify presentation and discussion.

Buzz groups are made up of two to four people who work together for a short time to complete a task, discuss a topic, or solve a problem. Buzz groups get their name from two characteristics of their activity: there is generally quite a noisy buzz in the room, and working in this way sets ideas buzzing. In buzz groups, consider the following question.

2. What elements of the ACP framework would be desirable in your organization and what would be most readily applicable? What would not work easily and why not?

4 The Public Utilities Board in Singapore

4.1 Introduction

Singapore is a rich city state of 4.4 million people with a public utility that provides exemplary water and sanitation services.[3] Although Singapore gets significant rainfall (2,400 millimeters a year), water is a rather scarce and precious commodity because of the limited storage space available in such a small state. Singapore is dependent upon water imported from neighboring Malaysia. The country has consequently promoted a wide range of supply augmentation and demand management strategies including desalination, wastewater reuse, and a carefully designed system of tariffs and taxes.

No matter which performance indicators are used, the Public Utility Board (PUB) invariably appears in the top 5 percent of all the urban water utilities of the world in terms of its performance.

- The entire population has access to drinking water and sanitation.
- The entire water supply system, from water works to consumers, is 100 percent metered.
- Unaccounted-for water represented 5.18 percent of total production (in 2004).
- The number of accounts served per employee is less than 400.
- Monthly bill collection efficiency is 99 percent.

At World Water Week in August 2007, the PUB received the 2007 Stockholm Industry Water Award.

4.2 Managing People

The overall governance of the water supply and wastewater management systems in Singapore is considered exemplary in its performance, transparency, and accountability. An institution can only be as efficient as its management and the staff that work for it, and the overall social, political, and legal environment within which it operates. In terms of human resources, the PUB has some unique management features that make it stand out among its Asian counterparts.

In the vast majority of Asian water utilities, service providers typically have only limited say in matters of staff recruitment and remuneration. Consequently, the utilities are rife with problems.

- Staff, including senior managers, are often selected because of their political connections rather than their management abilities or technical skills.
- Managers often do not have the skill to manage, even if they had autonomy and authority to manage, which often they do not.

4

- Water utilities are overstaffed, primarily because of political interference and nepotism. Unions are very strong and generally well connected politically.
- Accordingly, downsizing is a difficult task because of strong union opposition and explicit or implicit political support. Overstaffing ensures low productivity and low staff morale.
- Utilities are not allowed to pay their professional staff members the going market rates for remuneration, which sometimes can be two or three times higher. This means that they are unable to attract and retain high quality staff. Many staff members moonlight to obtain extra income, and corruption is rife in nearly all levels.
- Utilities are dominated by engineers, and the career structure available for other disciplines like accountants, administrators, social scientists, and information technologists is somewhat limited. This is another disincentive for nonengineers to join.
- Poor management, overstaffing, and promotions based on seniority or political connections ensure that it is very difficult to recruit good staff, and if some do join, it is equally difficult to retain them because of lack of job satisfaction, poor working environment, and absence of incentives for good performance.

The PUB has overcome these and other related constraints through a competitive remuneration, incentives, and benefits package. The salary and benefit package is generally benchmarked against the city's civil service rates, which, in turn, are benchmarked against the prevailing market. The PUB package provides strong performance incentives that are commensurate with the prevailing pay packages for the private sector. In addition, its pro-family policies, commitment to train its staff for their professional and personal development, and its practice of rewarding good performers all ensure good organizational performance and development. Consequently, its overall performance has become one of the best in the world.

4.3 Corruption

Corruption is endemic in most Asian utilities. However, it is not an issue at the PUB, which emphasizes staff integrity as a key organizational requirement. The utility has taken measures to prevent corruption by training staff in understanding and following the its code of governance and code of conduct, by implementing effective internal control processes, by conducting regular audits, and by imposing strong and immediate sanctions on those who prove to be corrupt. Staff members are required to make annual declarations on their assets, investments, and level of indebtedness.

Complaints of corruption are promptly investigated and reported to Singapore's Corrupt Practices Investigation Bureau. The PUB benefits from the overall environment in Singapore where there are strong anticorruption laws at the national level with appropriate sanctions that are regularly implemented. In addition, in recent decades, the government has consistently shown a strong political will to curb all forms of corruption and take firm actions against all and any form of corruption.

With a good remuneration package, a functional institution, and a strong anticorruption culture, corruption is not considered an issue at the PUB.

4.4   External and Internal Accountability

Clear and measurable targets are set for drinking water quality, customer service, and financial performance. An overview of the financial performance is published in the annual report, which is submitted to the government. Data on drinking water quality parameters can be found on the PUB Web site. The provisions of the Public Utilities Act for PUB's financial performance state that its total revenues must be sufficient to meet its obligations, including depreciation and interest on capital and a reasonable proportion of the cost of infrastructure development.

The main lines of accountability for the PUB are:

- To its owner: The PUB reports monthly to the Ministry of the Environment. The senior management of the utility (the board of directors and relevant department directors) meets with its parent ministry at least once a month on overall policy information and coordination. The PUB submits its financial reports annually to the government.
- To its regulators: The National Environmental Agency monitors drinking water quality, and its pollution control division monitors compliance with treated wastewater standards. The PUB reports to the Service Improvement Unit of the Public Services in the 21st Century Committee on a quarterly basis concerning its service quality levels.
- To financial institutions: Because the PUB currently has no debt, it is not accountable to outside financing agencies.
- To customer organizations and nongovernmental special interest groups: There is no organization representing the PUB's customers. However, the board of directors is made up of a broad spectrum of stakeholders.

The tariff structure and the setting of water tariffs and connections fees are proposed by the PUB and subject to approval from the government. Recommendations for capital sourcing are proposed by the utility and subject to approval by the Minister of the Environment.

The PUB financial manual stipulates the procurement ceilings. Expenditure of S$29,000 (Singapore dollars) to S$5.8 million requires Tender Committee B approval (this committee is made up of the chairman and two department directors). Expenditure in excess of S$5.8 million requires approval from Tender Committee A (made up of the chairman and two board members).

4

Hiring is decided by a recruitment committee convened by the human resources department. The committee is headed by a board member for the hiring of senior positions and by a senior staff member for other positions. Termination of service during the probationary period is decided by the head of department or by the chief executive officer if it involves senior positions. A promotion committee convened by the human resources department decides on the promotion of staff members. The committee is headed by a board member for promotion to senior positions, by the chief executive officer for the other professional grades, and by a head of department for the nonprofessional grades.

Decisions pertaining to the termination of service provision to defaulting customers are normally made at the department level. The way in which customer complaints are dealt with is also decided at the department level, guided by quality service standards.

A performance measurement system is in place to track the PUB's performance through key performance indicators. Employee performance is evaluated yearly through a staff appraisal process. During this process, the staff member is evaluated on indicators relating to achieved results, planning and organization, learning orientation, communication, team building, ability to lead change, and other similar criteria. Employees may be rewarded with performance bonuses or promotions. The promotion and the performance bonus evaluation processes are held yearly.

To manage poor performers, a performance review process is in place. In this process, an employee is counseled by the supervisor or union and advised on how to improve performance. If adverse performance persists, dismissal is an option.

4.5   Orientation to the Market and Customers

More than 25 percent of the PUB's operating budget is outsourced. In 2002, some 150 tenders and 170 quotes were let and requested, respectively, amounting to S$130 million. Contracts are let for a variety of services, including building construction, consultancy services, pipe laying, supplies, cleaning, security, information technology maintenance, and plant and equipment maintenance. The PUB follows public procurement rules.

The PUB has recently embarked on two benchmarking exercises on customer relations management and people management with the Public Service Center for Organizational Excellence, which is responsible for setting national norms. The utility is looking into market testing some of its operations and services.

The PUB draws funds for its operating and capital needs from the sale of potable water to its customers. It adopts a customer-focused approach to ensure customer satisfaction in all areas. The utility actively seeks the opinions of its customers through customer satisfaction surveys conducted every three years, regular dialogue sessions, and feedback forms. This continuous collection of feedback is aimed at providing a better understanding of customers' needs and expectations. The feedback is also used to identify areas for improvement in the PUB's operations. Emphasis is placed on the selection and training of frontline staff who come into direct contact with customers. They are specially trained in the areas of listening skills and service excellence.

The PUB operates a one-stop, 24-hour contact center (PUB One) for customers. Customers can make general inquiries and reports or provide feedback through this contact center through a variety of channels: telephone, e-mail, fax, SMS (Short Message Service), and VoIP (Voice over Internet Protocol). A public suggestion scheme has been introduced to reward customers for suggestions that improve PUB's services.

Activity 4.4

Effective Governance of Utilities

1. Which of the staffing problems identified in the management of the utilities apply to your own country? Can you identify other human resource problems?

2. What are the likely key success factors in the PUB operation that prevent corruption in the business of the utility?

3. What constraints would you face if you tried to replicate the transparency and anticorruption strategies used by the PUB in your own situation?

4

Notes

1. This case study is taken from Asian Development Bank (2006).
2. This case study is taken from www.pancanal.com/esp/general/transparencia/index.html.
3. This case study is taken from Tortajada (2006). For further information, see www.pub.gov.sg/home/index.aspx. The second part of this case is taken from Baietti, Kingdom, and van Ginneken (2006).

4

Action Planning to Address Corruption and Improve Transparency, Accountability, and Access to Information in the Water Sector

Donal O'Leary
María González de Asís

Contents

1   Introduction

1.1   Welcome

To make a difference in combating corruption in the water supply and sanitation (WSS) sector, it is necessary to move from analysis to action. In doing this, we should keep in mind that corruption is only one aspect of poor governance and that the overall objective is to improve the operation of WSS utilities for the benefit of the general good.

The previous four modules have discussed the nature of corruption in the water sector and have laid out a suite of comprehensive tools to diagnose and combat it. Numerous case studies, drawing on worldwide experience in the WSS and other sectors, have been used to reinforce the messages in these modules. This module will help you synthesize and apply the principal messages of the previous modules in working to improve governance and combat corruption in your own water supply and sanitation sector.

1.2   Goals of the Module

Based on the information provided in modules 1–4, this module sets out the major elements of an action plan to address corruption and improve transparency, accountability, and access to information in the different organizations involved in the WSS sector. It is focused on Honduras and Nicaragua but will be of use for developing programs elsewhere. The roles that can be played by different stakeholders, including the central government, municipalities, the private sector, consumers, and civil society, are also outlined.

1.3   Learning Objectives

At the conclusion of the module, the reader will be able to identify:

- The different elements of an action plan to diagnose and address corruption and improve transparency, accountability, and information in the WSS sector
- The elements of the action plan that are relevant to the different organizations involved in the WSS sector
- The roles that can be played by different actors in the water sector to address corruption and improve transparency, accountability, and information in the sector

1.4   Outline of the Module

The reader should complete modules 1 through 4 before studying module 5. Section 2 of this module provides the context for the exercise that follows through

5

a brief description of the WSS sector in Honduras and Nicaragua. Section 3 lays out the steps to develop a time-bound monitorable action plan aimed at addressing corruption and improving transparency, accountability, and information in the water sector. The module concludes with section 4, which provides you an opportunity to review the material and learning points.

1.5   Before We Start

Before starting, we would like you, the participant, to reflect on the situation in your own organization or in a water utility familiar to you by answering the following questions in activity 5.1.

Activity 5.1

Please answer the following questions:

Question 1: Are you aware of any corrupt activities within this organization?

Question 2: Are you aware of any anticorruption measures being taken by the utility?

Question 3: How effective were these measures?

5

2 Context

For the sake of completeness, this section is repeated from module 1. If you have recently completed module 1, you can skip this section.

2.1 Introduction

Major progress has been achieved in expanding water and sanitation services to the urban and, to a lesser extent, rural areas in Honduras and Nicaragua. However, the sector is still experiencing governance problems that are reflected in poor service quality. To address these problems, the two governments have initiated reforms of sector institutions that are proceeding at varying speeds. To provide a context for the subsequent discussion of corruption, this section gives a brief overview of the emerging institutional structure in the two countries and highlights some of the more critical performance issues.[1]

2.2 Sector Organization

The sector structure in **Honduras** is still being transformed after the approval of a new Water Framework Law (*Ley Marco el Sector Agua Potable y Saneamiento*) in 2003. The implementation of the sector reform is guided by the National Water and Sanitation Modernization Plan (*Plan Estratégico de Modernización del Sector Agua Potable y Saneamiento,* or PEMAPS) prepared in 2005.

The National Water and Sewerage Service (*Servicio Autónomo Nacional de Acueductos y Alcantarillados,* or SANAA) traditionally operated about half the urban water and sewer systems, with the balance managed by municipalities. The 2003 law calls for the transfer of assets and the responsibility for operating more than 30 water and sewerage systems from SANAA to the municipalities by October 2008. SANAA's role will change from being an operator to becoming a technical assistance agency supporting the municipally owned utilities. Policy making will rest with the National Water and Sanitation Council (*Consejo Nacional de Agua Potable y Saneamiento,* or CONASA), and a new regulatory authority for water and sanitation (*Ente Regulador de los Servicios de Agua Potable y Saneamiento*, or ERSAPS) has been created.

All urban water systems are publicly operated, except in San Pedro Sula, where the city has granted a concession contract to a private company, and in Puerto Cortés, where the city has created a mixed (private-public) company. A couple of municipalities have also entered into lease arrangements with private operators. Under the decentralization scheme, it is anticipated that municipalities will establish autonomous operators for the water systems. PEMAPS stresses the need for strengthened regulation, improved governance, and transparency.

In principle, municipalities are responsible for providing water and sanitation services in **Nicaragua**. However, in practice only 26 smaller municipalities actually

5

provide these services. Instead, most cities and towns are served by the national water and sewerage company (*Empresa Nicaragüense de Acueductos y Alcantarillados,* or ENACAL), which operates 147 separate systems. There are also three departmental water companies (in Jinotega, Matagalpa, and Rio Blanco) that together administer nearly 20 systems through management contracts with private companies. Some 5,000 rural water supply systems are run by community organizations with support from FISE, the Emergency Social Investment Fund.

Under reforms initiated with new legislation in 1998, the National Water and Sewerage Commission (*Comisión Nacional de Agua Potable y Alcantarillado Sanitario,* or CONAPAS) is in charge of policy making and sector planning, and the Nicaraguan Water Supply and Sewerage Institute (*Instituto Nicaragüense de Acueductos y Alcantarillados,* or INAA) regulates the sector through concession agreements.

As noted above, there is a modest amount of private sector participation in water supply and sanitation. However, in 2003, the legislature severely limited the scope for the further involvement of the private sector. It suspended "the awarding of all concessions to private individuals for operation of ENACAL's facilities and assets, or the awarding of management contracts to private individuals" and it changed ENACAL's status from a "state-owned business" to a "state-owned public utility."

2.3 Access and Service Quality

Although both Honduras and Nicaragua have made major progress in extending water services to the urban population, rural water supply remains more problematic, as does sanitation in both urban and rural areas (table 5.1). While the service levels (official access) lag behind those of richer countries in the region, they compare well with service levels in Asia and Sub-Saharan Africa.

The quality of service, however, is quite poor. Water is rationed in most **Honduran** cities under SANAA management and, according to the World Bank, water is supplied two times a week or even less in the summer. In 2000, according to the World Health Organization (WHO-UNICEF 2007), 98 percent of water systems in Honduras provided water on an intermittent basis for an average duration of 6 hours per day. Drinking water was being disinfected in only 51 percent of urban water systems, and only 3 percent of the collected wastewater was being treated, which led to pollution.

Water supply in roughly half of the localities monitored by INAA in **Nicaragua** is not continuous, and the share is higher in the summer. Urban drinking water quality generally meets WHO standards. It is also estimated that 42 percent of collected wastewater is treated (but few households are connected to a sewerage system).

2.4 Operating Efficiency

Nonrevenue water—or water that is not paid for—is estimated at 50 percent in Tegucigalpa, the Honduran capital, and 43 percent in San Pedro Sula.[2] In the early

5

Table 5.1: Share of Population with Access to Water and Sanitation Services in Latin America, 2004
(Percent unless otherwise indicated)

Country	Income 2004 GNI per capita (US$)	Water Urban Improved supply	Water Urban House connection	Water Rural Improved supply	Water Rural House connection	Sanitation Urban Improved facilities	Sanitation Urban House connection	Sanitation Rural Improved facilities	Sanitation Rural House connection
Haiti	400	52	24	56	3	57	0	13	0
Nicaragua	**830**	**90**	**84**	**63**	**27**	**56**	**22**	**34**	**0**
Bolivia	960	95	90	68	44	60	39	22	2
Honduras	**1,040**	**95**	**91**	**81**	**62**	**87**	**66**	**54**	**11**
Paraguay	1,140	99	82	68	25	94	16	61	0
Cuba	—	95	82	78	49	99	50	95	25
Colombia	2,020	99	96	71	51	96	90	54	20
Dominican Republic	2,100	97	92	91	62	81	65	73	27
Guatemala	2,190	99	89	92	65	90	68	82	17
Ecuador	2,210	97	82	89	45	94	62	82	16
El Salvador	2,320	94	81	70	38	77	63	39	2

(continued)

5

Table 5.1: Share of Population with Access to Water and Sanitation Services in Latin America, 2004 (Percent unless otherwise indicated) (*Continued*)

Country	Income 2004 GNI per capita (US$)	Water Urban Improved supply	Water Urban House connection	Water Rural Improved supply	Water Rural House connection	Sanitation Urban Improved facilities	Sanitation Urban House connection	Sanitation Rural Improved facilities	Sanitation Rural House connection
Peru	2,360	89	82	65	39	74	67	32	7
Brazil	3,000	96	91	57	17	83	53	37	5
Jamaica	3,300	98	92	88	46	91	31	69	2
Argentina	3,580	98	83	80	45	92	48	83	5
Uruguay	3,900	100	97	100	84	100	81	99	42
Venezuela, R. B. de	4,030	85	84	70	61	71	61	48	14
Panama	4,210	99	96	79	72	89	58	51	1
Costa Rica	4,470	100	99	92	81	89	48	97	1
Chile	5,220	100	99	58	38	95	89	62	5
Mexico	6,790	100	96	87	72	91	80	41	16
Trinidad and Tobago	8,730	92	80	88	67	100	19	100	—

Sources: Income data from World Development Indicators 2006; Water and sanitation data from WHO-UNICEF Joint Monitoring Programme (http://www.wssinfo.org/).

Note: —. Data are not available.

5

2000s, the water system in Tegucigalpa had more than 9 employees per 1,000 connections, which is a high number.

In Nicaragua it is estimated that 18 percent of the connections are illegal and 56 percent of the water supplied goes unbilled. ENACAL has 6.5 employees per 1,000 customers, which is nearly double the number regarded as acceptable for a company of this type (that is, 3 or 4 employees per 1,000 customers).

2.5 Tariffs and Financial Performance

Water and sewerage tariffs in Honduras are low, especially in municipal systems, which indicates that tariff setting in municipalities is prone to "political capture."[3] Tariffs barely cover operation and maintenance costs and subsidies are generally poorly targeted.

Tariffs charged by ENACAL in Nicaragua are high in relation to incomes. Still, the company is in poor financial health due to the operating problems discussed above. This has a serious impact on both new investments and operation and maintenance.

2.6 Governance

The above analysis indicates that the water utilities in both Honduras and Nicaragua face some serious governance problems. No surveys have been carried out that directly measure and describe corruption in the water sector in the two countries. However, both the World Bank's *Governance Indicators 2006* (especially the variable measuring a country's "Control of Corruption") and Transparency International's *Corruption Perception Index 2006* indicate that public administration in general is affected by corruption in these countries. On both scores, Honduras and Nicaragua perform well below average for the surveyed countries (table 5.2).

Table 5.2: Indicators of Corruption in Honduras and Nicaragua

Country	World Bank		Transparency International	
	Index	Rank	Index	Rank
Best country	2.49	1	9.6	1
Nicaragua	−0.62	133	2.6	111
Honduras	−0.67	140	2.5	121
Worst country	−1.79	204	1.8	163

Source: World Bank: Control of Corruption (range from −2.50 to +2.50) from http://www.worldbank.org. Transparency International: Corruption Perception Index (range from 1 to 10) from http://www.transparency.org.

Table 5.3: Corruption in Public Utilities in Honduras

Responses from enterprises and consumers regarding key public utilities	Percentage of respondents receiving high quality of service	Percentage of respondents who were made to feel that bribes were necessary	Amount paid in unofficial payments, (Lempira)	Percentage of respondents who did not make formal complaints because they believed it would not make a
Enterprises				
Phone installation	35	17	3,319	31
Electric connection	44	7	2,506	31
Water and sewerage	49	5	650	30
Consumers				
Phone installation	37	8	706	25
Electric connection	39	6	339	21
Water and sewerage	42	5	496	20

Source: World Bank Institute.

A diagnostic survey of governance and anticorruption in Honduras undertaken by the World Bank Institute (2002)[4] indicates that corruption is common in public utilities, including those in the water and sanitation sector (table 5.3). The same survey also shows that corruption is common in public sector procurement. More than one-third of private sector firms interviewed believed that corruption was frequent in public procurement and estimated that the bribes were around 12 percent of the contract value.

5

3 Developing an Action Plan

3.1 Introduction

In developing an action plan to address corruption in the water sector, we should bear a few things in mind.

First, the exercise focuses on addressing corruption issues in municipal water utilities as well as in the operation of "informal" or community systems to meet the WSS needs of consumers who are not hooked up to the water utility network.

Second, a necessary condition for its success is political will, that is, unequivocal and open support at the highest levels of all organizations affecting the success of such a plan. This commitment can be strengthened through the formation of a broad-based steering committee to provide strategic direction and review progress.

Third, there must be a commitment to a multistakeholder approach in developing, implementing, and monitoring the action plan. Ideally, the following stakeholders should be represented: the central government line ministry; the regulator; the municipality; the water utility/operator; the private sector; the consumers; and civil society. A stakeholder analysis can be used to understand each player's source of legitimacy, roles, responsibilities, and potential contribution to combating corruption. This analysis can also be useful in determining how best to engage vulnerable groups, such as the urban poor, in the process.

One could use the analogy of a football team to explain the roles involved in developing this plan, with specific actors taking on the roles of owner, coach, captain, other players, the referee, vendors, and the paying spectators. Once the team takes to the field, it will need to take into account the rules of the "game" as interpreted by the referee and be constantly nimble enough to manage the (inevitably) changing tactics of its opponents.

A **facilitator** could be used to help expedite the preparation of the action plan.

Fourth, one of the key initial steps will involve undertaking an initial diagnosis of the situation as well as establishing a baseline of data against which the impact of the program can be measured. To establish credibility, it is important to engage a neutral and respected organization (such as a university, a nongovernmental organization, or a management institution) to undertake this work. Unequivocal interim and final indicators of success need to be adopted that can be readily monitored and reported on by the media. To generate "buy-in" to the process, such indicators should be developed jointly and agreed on by all the key stakeholders.

Finally, to assure sustainability over the long haul in fighting corruption, it is very important to identify "champions" who will keep pushing for the necessary reforms, especially in those periods when political support may be perceived to be weakening.[5]

5

3.2 Step-by-Step Approach

Following is an exercise in developing an anticorruption action plan, which is broken into six subactivities.

Activity 5.2

Complete and Prioritize the Water Sector Corruption Risk Matrix

Please complete and prioritize the Corruption Risk Matrix below (based on the risk map presented in module 1) for the water agency you are most familiar with. The activities with the highest perceived risk should be assigned the number 1, with lower-risk activities scored from 2 to 5 according to the level of risk. An additional row has been included for comments. These could cover additional interactions not covered by the table including private-private interactions (such as bid rigging) and specialized supply contracts, including water treatment technology and information technology equipment. As part of this exercise, a judgment needs to be made about whether the institution is experiencing a limited amount of corruption by individuals or whether corruption is pervasive throughout the institution.

Activity	Public-public	Public-private	Public-consumer
Policy making			
Regulation			
Planning and budgeting			
Donor financing			
Fiscal transfers			
Management and program design			
Tendering and procurement			
Construction			
Operations and maintenance			
Payment (for services)			
Comments			

5

Activity 5.3

Select and Use the Diagnostic Tools

Using the results of activity 5.2, the next step is to select and use the diagnostic tool(s) described in detail in sections 5 and 6 of module 2 to further refine the analysis of activity 5.2. This could be carried out in the following sequence:

First, review the levels of **warning signals** (red flags) of certain performance indicators such as unaccounted-for water and the number of staff per 1,000 connections.

Second, in those cases, where the warning signals are at unacceptable levels, decide whether a **quick analysis** or a **detailed diagnosis** is needed. A quick analysis is appropriate in smaller and medium utilities, where financial resources are limited and where the water utility's staff and customers have a good understanding of the quality of service and potential governance problems.

Where a detailed diagnosis is needed, choose and use a pairing of an **external diagnostic tool** and an **internal diagnostic tool**. We suggest you choose a **pair** of external and internal diagnostic tools based on analyses of their strengths and weaknesses as well as on the perception of pervasiveness of corruption in the organization. Note that the Utility Checklist and the Vulnerability Assessment should be used in those situations where corruption is perceived to be limited to just a few individuals, but the Performance Benchmarking and the Public Record of Operation and Finance (PROOF) are more appropriate in situations where corruption is perceived to be pervasive throughout the organization. Space is also provided at the end of the table for comment.

Diagnostic tools	Strengths	Weaknesses	Choose (Yes/No)
External			
Corruption survey			
Planning and budgeting			
Donor financing			
Citizen report card			
Participatory corruption assessment			
Hybrid module			
Internal			
Utility checklist			
Vulnerability assessment			
Performance benchmarking			
PROOF			
Comments			

5

Activity 5.4

Identification of Impact Indicators and Their Baseline Values

A key output of the diagnosis stage is the identification of impact indicators and their baseline values. It will be up to the stakeholders to agree on ultimate and intermediate targets to be achieved within an agreed time frame. Where necessary, include any assumptions used in determining the target values, such as budgetary requirements to increase coverage area. A list of possible indicators is included on the worksheet for your information. Space is also provided at the end of the table for comments.

Impact indicators (units)	Baseline value	Interim benchmark	Target value	Assumption (s)	Target achieved? (Yes/No)
Number of staff per 1,000 connections					
Production capacity (cubic meters/day)					
Nonrevenue water (% of total)					
Number of people without access to WSS services					
Coverage area (%)					
Metered coverage (%)					
Supply duration (hrs/day)					
Collection ratio (%)					
Revenue (in US$)					
Impact of integrity pact (US$ saved)					
Freedom of information impact (number of complaints received and processed)					
Comments					

5

Activity 5.5

Identification of Anticorruption Tools in the Water Sector

Anticorruption tools are classified according to the TI and UN-Habitat (2004) scheme: access to information; ethics and integrity; and institutional reform and oversight. They should be selected in accord with the impact indicators identified in activity 5.4, which could reflect activities at the sectoral, institutional, and project levels. Altogether, 17 tools have been featured. Of these, one is described in module 2; 12 are described in module 3; and a further 4 are featured in module 4.

Anticorruption tools in the water sector	Strengths	Weaknesses	Choose (Yes/No)
Access to information and public participation			
Meetings to discuss corruption in the water sector			
Regular meetings			
Access to information laws			
Communication between water utility and its consumers			
Complaints and ombudsman office			
Community participation			
Budgetary/expenditure			
Public expenditure tracking (PET)			
Promoting ethics, professionalism, and integrity			
Integrity pacts			
Business Principles for Countering Bribery (BPCB)			
International conventions			
Institutional reform/oversight			
Independent auditing			
Regulator(s)			
E-procurement			
Conflict of interest policies; codes of conduct			
Disclosure of income and assets			
Collaborative models for providing WSS to the poor			
Comments			

5

Activity 5.6

Developing a Detailed Anticorruption Action Plan

In this activity, we ask you to draw up a detailed, time-bound, monitorable action plan to address the corruption problems your organization faces. Put each tool you have selected in one of the three plan stages (planning, implementation, and reporting). Be sure to spell out the roles of the different players at each stage. If it is helpful, use the analogy of the football match in allocating roles. A series of questions are set out below for each stage.

Planning

Who is responsible for preparing the anticorruption action plan? Which stakeholders will be consulted in its preparation? How will the preparation of the plan be financed? Who is responsible for signing off on the plan?

The output at this stage will be a report setting out how each tool will be implemented, including a staffing plan, an activity chart, its implementation, and its budget as well as a description of the role of each pertinent stakeholder in its implementation.

Implementation

Who is responsible for obtaining full political commitment to support this initiative? Who is responsible for securing funding to support this initiative? Who is responsible for identifying the manager for this initiative?

Reporting

Who is responsible for reporting on this initiative? How will the reporting be undertaken? How often should the reporting be undertaken? How will the other stakeholders be informed and/or involved?

Activity 5.7

Monitoring and Reporting on the Status of the Action Plan

The focus of this activity is to ensure independent monitoring of and reporting on progress on each impact indicator. Draw up a monitoring and reporting plan based on your answers to the questions set out below.

Question 1: What are the impact indicators and their targeted quarterly values throughout the duration of this exercise?

Question 2: How will an independent evaluation of the levels of the impact indicators be carried out? Will you use an independent auditor? Will consumers and/or civil society have any role? How can the water utility/operator support the process?

Question 3: What steps will be taken to inform the consumers and the general public of the results anticorruption action plan? Through press conferences? Through providing information in the bills sent to consumers?

5

4   Concluding Activity

We hope that you have found the material in this module useful and that it has helped you develop ideas for taking action using concepts and tools presented in the earlier modules. To give you an opportunity to review the material, we suggest you complete the final activity below.

Activity 5.8

Question 1: What are the next steps in implementing the action plan?

Question 2: What are the risks related to the implementation of the action plan you have set out? How can they be addressed?

Question 3: Identify how the results of the monitoring program can be used to ensure that the action plan meets its objectives in a timely manner.

5

Notes

1. This section is based primarily on documents from the World Bank (2007) and Inter-American Development Bank (2006), augmented by the entries on water supply and sanitation in Honduras and Nicaragua from Wikipedia (http://en.wikipedia.org/wiki/Water_supply_and_sanitation_in_Honduras) and (http://en.wikipedia.org/wiki/Water_supply_and_sanitation_in_Nicaragua).

2. Nonrevenue water (NRW) comprises three components: physical (or real) losses, commercial (or apparent) losses, and unbilled authorized consumption. The World Bank database on water utility performance (IBNET, the International Benchmarking Network for Water and Sanitation Utilities, at www.ib-net.org) includes data from more than 900 utilities in 44 developing countries. The average figure for NRW levels in developing countries' utilities covered by IBNET is around 35 percent (Kingdom, Liemberger, and Marin 2006). In a well-managed water system, the NRW would normally be below 20 percent.

3. Under political capture, regulation becomes a tool of self-interest within government or the ruling elite.

4. Also available in Spanish. See World Bank Institute (2002).

5. Further information on the development of action plans can be obtained in González de Asís (2000).

5

References

Andvig, Jens Chr., and Odd-Helge Fjeldstad. 2000. *Research on Corruption: A Policy Oriented Survey*. Study commissioned by Norwegian Agency for Development cooperation. Chr. Michelsen Institute and Norwegian Institute of International Affairs. http://www.icgg.org/downloads/contribution07_andvig.pdf.

Asian Development Bank. 2007. "Phnom Penh Water Supply Authority: An Exemplary Water Utility in Asia." www.adb.org/water/actions/CAM/PPWSA.asp.

Baietti, A., W. Kingdom, and M. van Ginneken. 2006. "Characteristics of Well-Performing Public Water Utilities." Water Supply & Sanitation Working Notes 9. World Bank, Washington, DC. www.worldbank.org/watsan.

Bailey, Bruce B. 2003. "Synthesis of Lessons Learned of Donor Practices in Fighting Corruption." DAC Network on Governance.

Balcazar, Alma Rocio. 2006. A five-minute video about the Colombian integrity pact between pipe manufactures can be viewed at www.waterintegritynetwork.net/page/254 (in Spanish).

Banisar David. 2006. "Freedom of Information around the World 2006: A Global Survey of Access to Government Information Laws." Privacy International.

Bretas, Paulo Roberto Paixão. 1996. "Participative Budgeting in Belo Horizonte: Democratization and Citizenship." *Environment and Urbanisation* 8(1) (April).

Clarke, George R. G., and Lixin Colin Xu. 2002. "Ownership, Competition, and Corruption: Bribe Takers versus Bribe Payers." Policy Research Working Paper 2783. World Bank, Washington, DC.

Davis, Jennifer. 2004. "Corruption in Public Service Delivery: Experience from South Asia's Water and Sanitation Sector." *World Development* 32 (1): 53–71.

Estache, Antonio, and Eugene Kouassai. 2002. "Sector Organization, Governance, and the Inefficiency of African Water Utilities." Policy Research Working Paper 2890. World Bank, Washington, DC.

González de Asís, María. 2000. "Coalition-Building to Fight Corruption." Draft Paper. World Bank Institute, Washington, DC.

———. 2005. "Reducing Corruption at the Local Level." World Bank, Washington, DC.

Inter-American Development Bank. 2006. "Loan Proposal: Nicaragua Water Supply and Sanitation Investment Program." October 4. Washington, DC.

Kaufmann, Daniel, Judit Montoriol-Garriga, and Francesca Recanatinil. 2005. "How Does Bribery Affect Public Service Delivery? Micro-Evidence from Service Users and Public Officials in Peru." http://papers.ssrn.com/sol3/papers.cfm?abstract_id=878358.

Kingdom, Bill, Roland Liemberger, and Philippe Marin. 2006. "The Challenge of Reducing Non-Revenue Water (NRW) in Developing Countries—How the Private Sector Can Help: A Look at Performance-Based Service Contracting." Water Supply and Sanitation Sector Board Discussion Paper 8. World Bank, Washington, DC.

Klitgaard, Robert. 1998. *Controlling Corruption.* Berkeley: University of California Press.

Klitgaard, Robert, Ronald MacLean-Abaroa, and H. Lindsey Parris. 2000. *Corrupt Cities: A Practical Guide to Cure and Prevention.* Washington, DC: Institute for Contemporary Studies (ICS) and the World Bank Institute.

Leautier, Frannie, Dani Kaufmann, and others. 2006. *Cities in a Globalizing World: Governance, Performance, and Sustainability.* Washington, DC: World Bank.

Menegat, R. 2002a. "Participatory Democracy in Porto Alegre, Brazil." PLA Notes 44. www.iied.org/NR/agbioliv/pla_notes/documents/plan_04402.pdf.

———. 2002b. "Participatory Democracy and Sustainable Development: Integrated Urban Environmental Management in Porto Alegre, Brazil." *Environment and Urbanisation* 14(2): 181–206.

OECD. 2007. "Integrity in Public Procurement: Good Practice from A to Z." Paris. http://www.oecd.org/dataoecd/43/36/38588964.pdf.

OSCE (Organization for Security and Co-operation in Europe). 2004. "Best Practices in Combating Corruption." Vienna. See http://www.osce.org/item/13568.html and http://www.osce.org/resources/.

Plummer, Janelle. 2008. "Water and Corruption: A Destructive Partnership." In *Global Corruption Report 2008—Corruption in the Water Sector.* Cambridge, U.K.: Cambrige University Press for Transparency International.

Plummer, Janelle, and Piers Cross. 2007. "Tackling Corruption in the Water and Sanitation Sector in Africa: Starting the Dialogue." In *The Many Faces of Corruption: Tracking Vulnerabilities at the Sector Level,* ed. Edgardo Campos and Sanjay Pradhan. Washington, DC: World Bank.

Reinikka, R., and N. Smith. 2004. "Public Expenditure Tracking Surveys in Education." www.unesco.org/iiep/PDF/pubs/Reinikka.pdf.

Ryan, Peter. 2006. "Citizens' Action for Accountability in Water and Sanitation. Water Aid." March. http://www.wateraid.org/documents/plugin_documents/wateraid_citizensweb.pdf.

Satyanand, P. M., and B. Malick. 2007. "Engaging with Citizens to Improve Services: Overview and Key Findings." Water and Sanitation Program-South Asia, New Delhi. www.wsp.org (in English only).

Schouten, T., and P. Moriarty. 2003. *Community Water, Community Management: From System to Service in Rural Areas.* Delft: IRC International Water and Sanitation Centre.

Shah, Anwar. 2006. "Corruption and Decentralized Public Governance." Policy Research Working Paper 3824. World Bank, Washington, DC.

Shordt, Kathleen, Laurent Stravato, and Cor Dietvorst. 2006. *About Corruption and Transparency in the Water and Sanitation Sector*. Delft: IRC International Water and Sanitation Centre.

Shordt, Kathleen, Lauren Stravato, and T. van Daalen. 2006. "Improved Transparency and Service Using Site Selection as a Tool." IRC International Water and Sanitation Centre, the Hague. www.waterintegritynetwork.net/page/395.

Shordt, Kathleen, Christine van Wijk, Francois Brikke, and Susanne Hesselbarth. 2004. *Monitoring Millennium Development Goals for Water and Sanitation: A Review of Experiences and Challenges*. Delft: IRC International Water and Sanitation Centre.

Smith, Terry. 2006. "The Potential for Participatory Budgeting in South Africa: A Case Study of the 'People's Budget' in Thekwini Municipality." CCS Grant Report: 1–37.

Sohail, M., and S. Cavill. 2007. *Accountability Arrangements to Combat Corruption: Case Study Synthesis Report and Case Study Survey Reports*. Leicestershire, U.K.: Loughboro University. www.wedc.lboro.ac.uk/wedc/publications/ (in English only).

Stålgren, P. 2006. "Corruption in the Water Sector: Causes, Consequences and Potential Reform." Swedish Water House Policy Brief 4. Stockholm International Water Institute, Stockholm.

Stoupy, O., and S. Sugden. 2003. "Halving the Proportion of People without Access to Safe Water by 2015. A Malawian Perspective." Part 2: "New Indicators for the Millennium Development Goal." WaterAid, Malawi.

Swain, Biraj, James Wicken, and Peter Ryan. 2006. "Citizens' Action: How Bridging the Accountability Gap Leads to Improved Services". Paper presented at the 32nd WEDC International Conference, Colombo, Sri Lanka. www.lboro.ac.uk/wedc.

Swardt, C. de. 2005. *Global Corruption Report 2005–6: Key Developments in Corruption across Countries*. Berlin: Transparency International.

TI (Transparency International). 2001. *Corruption in Kenya: Findings of an Urban Bribery Survey*. http://www.tikenya.org/documents/urban_bribery_index.doc.

———. January 2003. *Global Corruption Report 2002: Special Focus – Access to Information*. London: Pluto Press.

———. 2005a. "Business Principles for Countering Bribery." Transparency International, Berlin.

———. 2005b. *Global Corruption Report 2005: Special Focus: Corruption in Construction and Post-Conflict Reconstruction*. London: Pluto Press.

———. 2006a. *Corruption Perception Index 2006*. http://www.transparency.org/policy_research/surveys_indices/cpi/2006.

———. 2006b. *Handbook for Curbing Corruption in Public Procurement*. Berlin: Transparency International.

———. 2006c. "Using the Right to Information as an Anti-Corruption Tool." www.transparency.org.

Tortajada, C. 2006. "Water Management in Singapore." *Water Resources Development* 22 (2): 227–40.

Transparency International Bangladesh. 1997. "Survey on Corruption in Bangladesh—Executive Summary." www.ti-bangladesh.org/evidence/evidence-app3b.pdf.

———. 2000. "Corruption in Public Sector Departments: Its Manifestations, Causes, and Suggested Remedies." Report prepared for the World Bank. http://www.ti-bangladesh.org/index.php?page_id=331.

———. 2005. "Corruption in Bangladesh: A Household Survey—Executive Summary." http://www.ti-bangladesh.org/HH%20Survey/Household%20Survey%20-%202005.pdf.

TI and UN-HABITAT. 2004. *Tools to Support Transparency in Local Governance.* www.transparency.org/tools/e_toolkit/(in English only).

Trémolet, Sophie, Rachel Cardone, Carmen da Silva, and Catarina Fonseca. 2007. "Innovations in Financing Urban Water and Sanitation. Background Paper, Rockefeller Foundation Urban Summit. June.

UNDP (United Nations Development Programme). 2006. *Human Development Report 2006—Beyond Scarcity: Power, Poverty and the Global Water Crisis.* New York: Palgrave Macmillan.

WHO (World Health Organization) and UNICEF (United Nations Children's Fund). 2007. "Joint Monitoring Programme (JMP) for Water Supply and Sanitation." http://www.wssinfo.org/en/welcome.html.

World Bank. 1998. "Helping Countries Combat Corruption: The Role of the World Bank." Poverty Reduction and Economic Management Network, World Bank, Washington, DC.

World Bank. 2006. *Governance Indicators 2006.* World Bank Institute. http://info.worldbank.org/governance/wgi/index.asp.

———. 2007. "Project Concept Note: Honduras Water and Sanitation Program." March 1. World Bank, Washington, DC.

World Bank and BEI (Bangladesh Enterprise Institute). 2003. *Improving the Investment Climate in Bangladesh.* Washington, DC: World Bank.

World Bank and UNDP (United Nations Development Programme). 2002. *Bangladesh: Financial Accountability for Good Governance.* Washington, DC: World Bank.

World Bank Institute. 2002. *Gobernabilidad y Anticorrupción en Honduras: Un Aporte para la Planificación de Acciones: Escuchando las Voces de Los Funcionarios Públicos, Empresas, y Usuarios de Servicios Públicos.* Análisis preparado por el a solicitud del Gobierno de Honduras para su discusión con el Consejo Nacional Anticorrupción 9 de Enero de 2002. http://info.worldbank.org/etools/docs/library/206680/Gobernabilidad_honduras.pdf.

———. 2002. "Governance and Anti-Corruption in Honduras: An Input for Action Planning." Analysis prepared at the request of the Government of Honduras for discussion with the Consejo Nacional Anti-corrupción. http://info.worldbank.org/etools/docs/library/206690/hon_gac.pdf.

———. 2003. "Pilot Municipal Anti-Corruption Course for Africa: Program Manual." Washington, DC.

———. 2004. "Reducción de la corrupción al nivel local; Curso de gestión urbana y municipal (educación a distancia)." Washington, DC.

Index

Boxes, figures, notes, and tables are indicated by *b*, *f*, *n*, and *t*, respectively.

www.ingramcontent.com/pod-product-compliance
Lightning Source LLC
Chambersburg PA
CBHW080238270326
41926CB00020B/4292